D0814210

GOOGLE DATA STUDIO
FOR BEGINNERS

START MAKING YOUR DATA ACTIONABLE

Grant Kemp

Gerry White

Apress®

Google Data Studio for Beginners: Start Making Your Data Actionable

Grant Kemp
London, UK

Gerry White
London, UK

ISBN-13 (pbk): 978-1-4842-5155-3
https://doi.org/10.1007/978-1-4842-5156-0

ISBN-13 (electronic): 978-1-4842-5156-0

Managing Director, Apress Media LLC: Welmoed Spahr
Acquisitions Editor: Shiva Ramachandran
Development Editor: Rita Fernando
Coordinating Editor: Rita Fernando

Cover designed by eStudioCalamar

Distributed to the book trade worldwide by Springer Science+Business Media New York, 1 New York Plaza, New York, NY 100043. Phone 1-800-SPRINGER, fax (201) 348-4505, e-mail orders-ny@springer-sbm.com, or visit www.springeronline.com. Apress Media, LLC is a California LLC and the sole member (owner) is Springer Science + Business Media Finance Inc (SSBM Finance Inc). SSBM Finance Inc is a **Delaware** corporation.

For information on translations, please e-mail booktranslations@springernature.com; for reprint, paperback, or audio rights, please e-mail bookpermissions@springernature.com.

Apress titles may be purchased in bulk for academic, corporate, or promotional use. eBook versions and licenses are also available for most titles. For more information, reference our Print and eBook Bulk Sales web page at http://www.apress.com/bulk-sales.

Any source code or other supplementary material referenced by the author in this book is available to readers on GitHub via the book's product page, located at www.apress.com/9781484251553. For more detailed information, please visit http://www.apress.com/source-code.

Printed on acid-free paper

Contents

About the Authors

Grant Kemp is a Data Studio specialist who regularly delivers trainings and speaks at conferences and meetups to share the transformative power that Data Studio can offer. He has helped a wide variety of companies from small retailers to multinationals to start using Data Studio.

Grant has over 18 years of experience in digital, starting out as a developer and working across multiple verticals including ecommerce, publishing, startups, and travel. He has deployed data solutions within a bevy of well-known companies, such as Dreams, Gap, Photobox, Missguided, Arsenal Football Club, and Virgin, among others.

Gerry White is an experienced digital marketer specializing in SEO and analytics, particularly focused on technical elements of site performance. He has worked for clients such as the BBC, McDonald's, Weight Watchers, BHS, Gordon Ramsay, and Premier Inn to name but a few.

About the Technical Reviewer

Alex Gonçalves is the author of *Social Media Analytics Strategy*, published by Apress, and has a background of over 10 years in analytics, performing analytical work as well as training individuals and organizations on analytics understanding, implementation, and strategy. Working with varied organizations in very different sectors – such as Autodesk, Samsung, Benefit Cosmetics, New York Botanical Garden, and Havas agencies, among so many others – Alex constructed a broad schema of knowledge around the specific needs and uses of analytics under varied contexts. Currently Alex is invested in an academic career researching human learning and motivation, among other topics. Finally, Alex also collaborates as a reviewer and editor on publications of books and papers, teaming up with authors toward producing the best possible product from the initial material. You can find Alex on LinkedIn at `www.linkedin.com/in/alexgoncal`.

Starting Your Data Studio Journey

Building Your First Data Studio Report

Questions are good. Questions allow us to see the beauty of life. Never stop asking questions.

—Peter Cawdron, *Hello World* (self-published, 2019)

Most people who are seen as "data experts" or "data gurus" may seem like they are "wizards" now, but they were all beginners once. One day, something happened that made them decide that they wanted to start experimenting with a data visualization tool. Maybe one weekend or one evening instead of slumping in front of the TV, they decided to fire up their laptop and start experimenting with a data tool. For them it would have been quite frustrating, as frankly the documentation is usually not great, and examples are often harder to understand. I found on my personal journey that I could sink hours into understanding various concepts and ideas even to do things that I felt

G. Kemp and G. White, *Google Data Studio for Beginners*,
https://doi.org/10.1007/978-1-4842-5156-0_1

were quite basic. I managed to find my way through, but not without a lot of stress and painful learning.

The good news is that we are going to do this journey together. In this chapter, I am going to take you through your first steps with Data Studio and work with you until you are ready to go out and hold your own.

So pull up a chair, make sure you have a nice drink next to you, and we are going to start experimenting with building your very first data report in Data Studio.

Starting Your Data Journey

I always say you don't find data but rather data finds you. I have worked with countless people across a wide variety of industries, each of whom was at different stages of data understanding and interest. Through my talks at conferences and meetups, I have had the pleasure of encountering people who are at the early stages of their interest in data. What I find most remarkable is just how many different backgrounds people come from. The one common thing they all share is that they found themselves at a challenge or a crossroad where they were stuck. They knew that there was a missing piece in their understanding and were seeking for data to fly in and be the superhero to save them.

I would love for data to be a magic bullet that can simply be switched on to solve problems; however, in reality, finding the data is just the beginning of the journey. For some people, they will get part way down the data road and discover they have enough and veer off to something else they find interesting. For others, they will be thrust forward into the multitude of wonderful new data job specializations that are sprouting weekly.

Data scientist, data engineer, artificial intelligence engineer are all new roles that were quite rare a few years ago but are becoming increasingly commonplace in most organizations. You could argue that to even be a reasonably effective digital marketing manager nowadays takes the level of data literacy that previously would have been required of a data analyst.

Whatever path you take (or not) the one standout fact that I have learned is that everyone can benefit from data visualization skills. It helps you to

- Tell stories with your data
- Take your audience on a journey with you
- Help your audience to understand your data and decide insights from it

On a personal level, I find that my brain just doesn't process a wall of numbers or text. I use data visualization as a method to explore and understand numbers. It's the difference between looking at a picture of a beach and feeling the sand between my fingers and hearing the sea wash up on the shore.

Data visualization brings the data to life and turns it into something real.

Business Attitudes to Data

One of the watershed discoveries that I have made in my career is that businesses are a lot like people when it comes to their use of data. The use of data varies drastically based on people's lives and experiences. Some people live their lives highly dependent on data, be it analyzing budgets, planning their future, or optimizing things like their bills, their commute, or even where they buy their groceries. Other people just go with the flow and rely on their personalities and instincts to help them get what they want in life.

When you start applying data to a business, you first need to understand the business' attitude to data. Within the company's culture, data is either a core driver of the business or an annoyance foisted upon others from those above with the sole mission to prevent others from doing what they want.

I think businesses are far too complicated and important entities not to rely on data for making decisions. There is only so far that a good brand or product will take you, before you need to start understanding the market and business goals to succeed. When I encounter people who scoff at data, I always put the point that data is never the priority for a business until it becomes *the* priority (and this happens incredibly rapidly and very suddenly when things stop going well).

When I was at university, I would sit through endless boring classes on stats and would hate that it was never practical, always theoretical. Had I known what applications it would have, I probably still wouldn't have enjoyed it as it wasn't real for me. I thoroughly wish that I would have had real-world examples to teach me the various aspects of the craft of data. As someone who learns by doing, this is the most effective way for me to understand anything, and that's what I have tried to do with this book.

So let's continue our journey into Data Studio together. I remember my journey fondly, and I hope I can shortcut some of the pain and share some of the positive experiences that I picked up as I took my various steps (and missteps) on the path to being comfortable with data.

Remember: You Are a Data Developer

When people hear the word "developer," they have several preconceived ideas about what that role means. The most common impression is that a developer is someone who is a genius and is able to process massive algorithms in their head in a computer-like way to produce code. This is a misguided idea. Developers are just regular people who have invested their time to learn a programming language and to understand the tools, techniques, nuances, and quirks of the technology they are using to produce results. Any professional job, be it a bricklayer, plumber, or indeed a developer, requires you to learn the tools of the trade to build what people need. A developer, by my definition, is simply someone who creates something out of nothing. A property developer takes a piece of land and produces homes for people to live in. A web developer produces apps or websites from nothing.

In most older, slower-moving organizations, the people who are responsible for creating dashboards in a business would have a job title that includes the term "analyst" or "data manager." If you work in one of these organizations, I want you to take a step out of the "analyst" box and start to think of yourself as being a "developer" and adopting a developer mindset when it comes to building out your data visualization reports.

In some more forward-thinking organizations, the job title given to those people who specialize in building dashboards or creating data pipelines is usually called BI (business intelligence) developers or data engineers.

By taking this first mental step forward and calling yourself a developer, you are bringing these three facts to the front of your mind:

1. As someone who builds data visualization systems that businesses rely on for progress and decision making, you are just as valuable as someone who makes an ecommerce store work correctly.

2. Your work can have the vital impact on a business to help them succeed.

3. With data development skills, comes the duty to have care that what you are developing is quality, accurate, and fit for its intended purpose.

When I have built data reports for businesses, I have benefited from using processes that are typically associated as being developer processes. Not because I try to make things more complicated but rather to focus on delivering trusted and maintainable data visualizations. The data developer roles are probably one of the most impactful roles that exists in a modern business, so it's up to us to do the work to get things right. So congratulations on entering the world of being a developer!

Getting Started: Hello World!

Whenever any developer starts a new programming language, they typically start off with writing a program that shows the words "Hello World" on the screen. Since you are all burgeoning data visualization developers, we are going to kick off our journey by using Data Studio to say hello to the world. We're going to create a simple report that pulls data from a CSV file.

▓ **Note** Google Studio is constantly evolving with new releases and features. As such, the screenshots in this book may look slightly different from what you see on your screen.

Step 1: Getting Setup with a Google Account

Data Studio is a free tool for anyone to use. In order to access Data Studio, you need to have a Google-enabled account. Good news! If you have a Google, Gmail, or a G Suite business account, you are ready to get started and you can skip to the next step.

For those of you who don't have a Google Account, you can visit this URL: https://accounts.google.com/signup/v2/webcreateaccount.

On this page you can decide if you want to sign up for a new Gmail.com address or to activate an existing email as a Google Account. Either type in the username you want to use for your new Gmail address or click "Use my current email address instead" as shown in Figure 1-1.

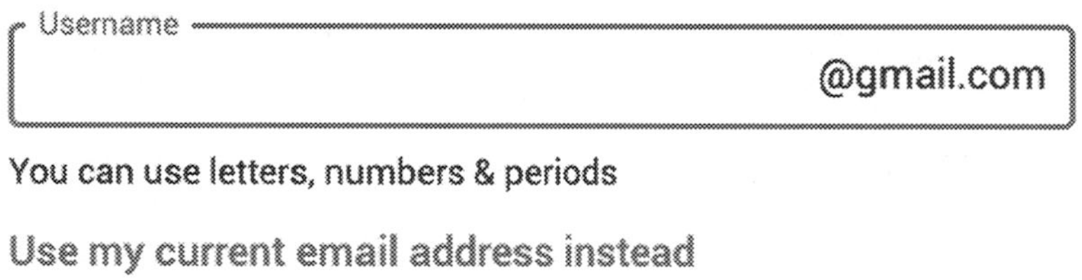

You can use letters, numbers & periods

Use my current email address instead

Figure 1-1. Setting up your new Google Account using a new Gmail address or an existing email address

Fill in the sign-up form with your details to continue.

Step 2: Open Up Data Studio

After you have a Google Account, head on over to `https://datastudio.google.com` to open up the Data Studio home screen.

The Data Studio home screen (Figure 1-2) is your launch pad into the world of Data Studio. Before we create our first dashboard, let's get familiar with the interface (Figure 1-3). It's made up of four key parts:

1. **The Template Gallery** gives you a quick way to get started with a new dashboard. You can either use a blank report or use a premade template that was created by the Data Studio team. To jump in with a premade template, press the "All Templates" button in the top right-hand corner.

2. **The Report Organizer area** is just below the template bar, and it is a set of organizational columns or filters that allow you to sort your reports based on

   - Name
   - Owner
   - Last opened

3. **The left menu** gives you a quick way to

   - Create new reports
   - Access reports that were created by others and shared with you
   - Access reports that you created or copied from others
   - View the trash for any reports you have deleted

4. **The search bar** at the top is a handy tool to let you search for reports by name.

Figure 1-2. Data Studio home page

Figure 1-3. Data Studio interface

Step 3: Create a Blank Report

To create a blank report, click the "Create" plus icon either in the upper left corner of the screen and select Report, or you click the plus icon in the template bar.

The Data Studio interface always looks quite intimidating at first glance. Don't worry. By the end of this book, you will know the interface like the back of your hand as you will have used all the features on it. For now, we will just run through the main core areas of the interface:

- The report title
- The sharing button
- The view/edit mode button
- The primary toolbar
- The reporting canvas
- The configuration panel

Note You might see an introduction screen to help you connect a data source. We'll get to that shortly, but for now, just close it so you can get to the main interface.

Step 3.1: Name Your Report

Click the text "Untitled report" in the upper left corner of the screen to rename your report. Choose something easy to understand, such as "Learning - Hello World." Then, press enter or click away from the name box.

It's to use good practice to keep your workspace tidy and you protect your data. One useful habit that I have learned is that I will always start a new report by renaming it from the default name that Data Studio gives it.

The best names are those that are memorable and include the terms that you are likely to search for when trying to find it in a list. Choosing a good name for a report is like writing a love letter to yourself in the future. You will feel so happy that you took the time to save yourself that extra time trying to find it. I can't tell you the amount of time that I have saved just by having a good report name.

My most common names that I use are

- For clients or brands I am working for, I use "Brand - Acquisition report" or BrandA - YouTube weekly report.

- If I am just creating a new report to play around with a feature, I use the term "Playground" to describe it – for example, "Calculated fields Playground."

- If I am just creating a temporary report to experiment with something, then the best possible name to use is simply "delete me" so it doesn't become another report to open when I am looking for something important.

Step 4: Creating Your First Data Connection

When we think about building a Data Studio report, it's a little like being a plumber. We want to make the water flow from the outside to our main sink where we can make use of it for washing, cooking, and drinking. To get this first report working, we need to make a connection to our main "data" water supply. In this case, we will be getting data from a CSV file.

Step 4.1: Download the Sample "CSV" File

We are going to connect directly to the most basic possible data source: a CSV file. For this example, we are going to use a simple text file that I created using a text editor. The file is in a standard format called a "comma separated values" file (or CSV for short). If there was a universal language for exchanging data between systems, then CSV would probably be it.

Download the sample CSV file (Ch1_ExampleCSV.csv) from the book's source code, available at www.apress.com/9781484251553.

Most platforms that people use will usually have support for exporting their data into CSV, so this exercise is going to be a technique that you are going to use over and over.

Step 4.2: Upload the CSV File to Data Studio

Now that we have got our file of data, we need to get it put into a form that Data Studio can access and manipulate. We do this by clicking the "Add data" button from the toolbar. You'll then see a long list of different types of data sources. At the time of print, we have

- **Google Connectors** – This includes common connectors such as Google Ads, Google Analytics, and Google Sheets. I find that these connectors are generally the most reliable as they have been built by the internal Google teams and are directly supported by Google.

- **Partner Connectors** – The partner connectors have been created by third-party companies and generally require you to pay for their use. These companies generally support the plug-ins directly; however, you may have mixed experiences if you rely on these for your reporting. You generally have to rely on the company that created them to make changes as you can't see the code for them. Not all of these connectors have large support teams and can take a long time to address any issues that you have. Having said that, some of these connectors are extremely powerful such as the very popular "Supermetrics" connector which lets you connect to Google Search Console, Facebook Insights, and even to payment providers such as Stripe.

- **Open Source Connectors** – These are data sources that are generally built by wider community members and are contributed to Data Studio for the benefit of others. The advantage of these is that you can directly see the code for them and thus can change them if it's relevant for your project. Most of these connectors are free; however, like other open source projects, you are reliant on other people to fix any issues.

The connector that we want to use right now is called "File Upload." Once you have selected it, you need to drag your CSV file into the upload box, or click the button to browse for the file on your computer.

You will be asked to Authorize access for Data Studio to upload this file to your Google Cloud Storage. We won't go into too much detail on Google Cloud Storage in this book, but you can think about it as being a secure hard disk somewhere on the Internet. Google does the hard work of hosting, securing, and protecting access to it so you don't have to worry about unauthorized people having access to it. Click Authorize and Accept to give Google Data Studio access to your storage.

Once you have approved Google to store your CSV file into Google Cloud Storage, we are finally ready to upload our data.

After a few seconds, the status of the file has changed from uploading to processing to uploaded.

You will see there are four status fields at the bottom of the page. These include

- The file name you uploaded
- When the file was uploaded
- Size of the file
- Status of the file

The Uploaded status has a handy little green dot next to it, which indicates that Google is happy with the data and you are good to continue connecting.

Once the file is done uploading, you have the option to rename the source by clicking the file name. You should always give your data sources an easy-to-understand name. This will help you identify it later on. A bit of tidying up front will save you a lot of pain at the end, especially when you have hundreds of data sources littering your data sources folder. For this exercise, let's call it "Hello World - CSV." See Figure 1-4.

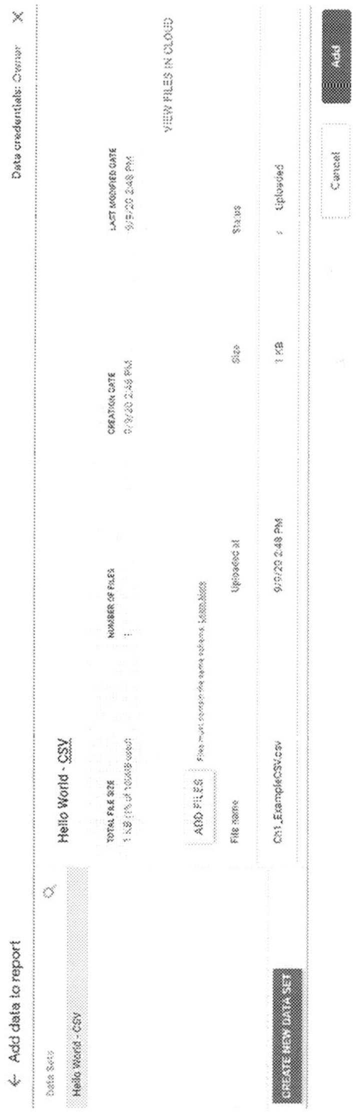

Figure 1-4. Uploading a CSV file data source

Selecting the blue "Add" button will let you select the CSV file to upload it to your account.

Step 4.3: Edit Connection and Authorize Access to Google Drive

You will see a little pop-up dialog about adding data to this report. For now, just go ahead and click "Add to Report." Google will request the permission to create a new Google Data Studio Report in your Google Drive.

▓ **Note** I personally find this step a little confusing, as I don't normally have to ask for permission to add a Google Doc or Google Sheet to my Google Drive. I am not sure why Google treats a Data Studio document any differently. I hope Google is able to remove it and make things simpler for users in the near future.

After the dialog box, you will return to the main Google Data Studio report editor screen. You can now celebrate as you have successfully created your first Data Studio connection. Now we can focus on showing the data on screen for the world to see your greeting.

Step 5: Display the Data from the CSV in Data Studio

Now we are ready to show off your data to the world. In the menu bar, you will see a button that says "Add a chart." When you click it, you can select from a multitude of different ways to visualize your data. We will go into these later on in more detail. For now, just select the first icon for a simple table. You will see a large rectangle appear on the screen with a small table icon in it. Data Studio is asking you where you would like to place this table. For now, just place it somewhere nice and tidy at the top of the page, where it is showing on the main canvas. See Figure 1-5.

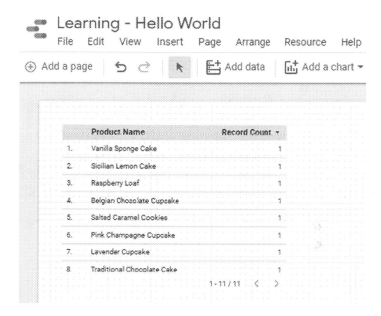

Figure 1-5. Our first table in Data Studio

As if by magic, you can stare as your first data table fades into existence. I am sure you have seen tables show up in many programs before; however, what makes Data Studio a game changer for data people is just how powerful and flexible that little table is. I like to think of it as being a jetpack, when you are in a road race. If you want to get ahead, just whack on those rockets to blast ahead of the rest of the pack.

Recap

In this chapter, you have started from zero and built up a foundation for Data Studio. We have taken some data that was held in a file and connected to it in Data Studio. We have used a table control to display it on screen so we can say hello to the world.

In the next chapter, you will get your first ever Data Studio assignment and build a basic dashboard for a small business.

Reporting

Manual Spreadsheets and Data Studio Reports

I never teach my pupils, I only attempt to provide the conditions in which they can learn.

—Albert Einstein, theoretical physicist

It looks like your first "Hello World" Data Studio report from Chapter 1 was just the start! What started as a personal practice session in building your first report has snowballed into your first dashboarding opportunity.

I have found that sharing what you are passionate about is always a good thing. If it's something that really excites you, then people with similar interests are naturally pulled toward you. You will also see opportunities materialize from places you weren't aware of before.

There is nothing more useful than learning a new technology by building a small project with it. If you can earn something for your learning projects, then you are in a very exclusive band of people and super lucky. In my case, I spent a lot of my spare time learning Data Studio, and I enjoyed it because I like the creative aspect of what Data Studio could do with my data. I would park out on my kitchen table after the kids went to sleep, and instead of slumping in front of the TV, I would try to build something cool in Data Studio.

© Grant Kemp, Gerry White 2021
G. Kemp and G. White, *Google Data Studio for Beginners*,
https://doi.org/10.1007/978-1-4842-5156-0_2

Tip   Take a few minutes to think about what kind of projects you would like to build with Data Studio. It could be something personal like plotting the distance you exercise in a week. It could be work related like automating your own reporting.

So, let's get started with our first little project and help our friend Hermann by making a sales report.

Helping Hermann's Cake and Coffee Shop

Hermann has been running his own coffee shop and bakery for the last two years and has built up quite the loyal following. Some of his most loyal customers even call him "Uncle H" as he's always around to share a story.

He used to be a motorbike mechanic in one of the older parts of the city but found that with all the tech companies moving in, he could make quite a lucrative business from selling them coffee and cakes. It turns out that techies really love having cool and unique places to hang out and work. Having spent most of his time hanging out with bikers, he has a lot of great stories which he would often entertain his patrons with over a nice hot drink and pastry.

Nowadays, the fervor which he used to devote to his motorcycle business has been transferred into his coffee shop. He focuses on having the finest organic and vegan food, and he is obsessed about tuning his coffee so that his coffee is ten times better than the shops down the road.

Figure 2-1. Hermann's coffee shop

How can you help him to understand a bit more about his business using your new Data Studio techniques (Figure 2-1)? He will pay you with free coffees and cupcakes for a month if you can help him to make more money.

▓ **Note**   It's OK to do work for free while you are learning. Once you have the skills though, don't expect to work for free. No one should ever be expected to give something for free in return for getting some "future" benefit or promotion. It is very rare for those opportunities to materialize into something lucrative. If someone wants your skills, they should pay you. When it comes to Hermann, we will let him off this time with the coffee and cakes offer after all it's a really good way to see what we can do with Data Studio and build that first project for our portfolio.

What Information Should You Show in Your Reports?

Whenever I meet a new client or business stakeholders, I always try and take the time to understand them with four key questions:

1. **What is their current goal?** What is their main goal for the business?

2. **What information do they need?** What information do they need to help drive this goal?

3. **What is their level of data expertise?** How sophisticated are they in using and understanding data?

4. **How should it be presented to them?** What's the best way to deliver data that is being clearly understood by the audience?

The first question is the most important one to understand. All people are trying to achieve something, so if you can make sure you have built alignment with them, it makes it easier for you to work together.

Identifying the exact information that will help them toward that goal is the next question. Typically most people will come to you with a set of requirements that they believe they need. It's up to you as the data person to understand those requirements and make sure that they are really helping them to achieve what they need. If they don't, you will find out really quickly because no one will look at your reports.

As you produce more reports, you will pick up experience with being able to filter out the "extra noise" and will be able to hear the various cues that indicate that the person is not sure of what they want or need and is just using their best guess. This will let you step in and guide them toward something more concrete and helpful.

The audience's data expertise is covered in the third question, and it's extremely difficult to gauge at the start of your working relationship with them. It's also a question that most people will very rarely answer truthfully up front. People are generally very defensive when you ask straight out about how much they know about data. Instead of asking people outright, there are much better techniques to get this information.

I have had people tell me that they use Google Analytics "all the time" yet, when I look at their Google Analytics accounts, I can see that the accounts are hardly used (and in many cases giving inaccurate data). It's not just at manager level, I have several experiences of working with extremely senior data directors who claimed to be experts, and I was surprised to find out that they didn't understand simple concepts such as user conversions or bounce rate. Just because they got to their position without knowing the fundamentals of data analytics doesn't mean you can't teach them something new. If you are dealing with people who are not very sophisticated in terms of their data experience, then being able to give them answers in a simple and clear way will be vital.

On the other hand, some users are extremely sophisticated, and if you give them simple answers and no detail, they will quickly become frustrated and not trust the results you are giving them.

The fourth question, "How should it be presented," is the most important one. The more you practice building dashboards, the more effective you will be at getting traction in your business.

Finding Out What Hermann Needs

Most of the world will make decisions by either guessing or using their gut. They will be either lucky or wrong.

—Suhail Doshi, the CEO of Mixpanel

When you start working with a new client or stakeholder, you need to try and understand what their need is. For Hermann's coffee shop, it's pretty clear. He has grown his shop organically by focusing on building relationships with customers and having a great quality set of products that people love. In the old days, that was enough, but nowadays, if you aren't using data to drive your business, then you are only seeing a "slice" of the picture.

Hermann's brief is to challenge you to use data to help him make more money. This is a pretty big ask and one we are unlikely to be able to fulfill. A lot of people just assume that waving the data "magic wand" on their business will suddenly materialize into money. Hermann is clearly one of these people. This is typically the part where you try and help them understand where data sits in the food chain.

Your answer should always be the same. You explain to Hermann that together we can use data to try understand better what's going on which can help inform his decisions. Sometimes that leads to more money being made, sometimes it doesn't.

Most business owners will look a little uncomfortable when you tell them that you might not be able to give them what they are expecting, but that's OK. It's better to make sure you manage people's expectations instead of trying to deliver something that's not realistic. The important thing is build up a good history of showing results and you will be able to use this to build trust early on with potential customers/internal teams.

The one guarantee that you can give anyone is that by making his data visible and easily accessible you can help to make smarter decisions in order to grow any business.

Back to your meeting with Hermann, you have a list of the four key questions you would like to ask him based on the previous section.

What Is His Current Goal?

Hermann tells you that he's looking to grow the amount of sales for the coffee shop. If he is able to do this, he will be able to open up a new coffee shop in one of the other tech districts in the city.

What Information Does He Need?

Hermann says that he currently logs the takings each day in a simple Excel document that he sends off to his accountant to enter into his accounts system. He then gets a report emailed to him at the end of the week with what the top selling products were. He uses this information to order his ingredients and plan for what he's going to sell the following week.

What Is His Level of Data Expertise?

Hermann admits to you that he's out of his depth when it comes to data. He knows what sells but he has no idea about using some of the common tools in spreadsheets. He says that he's quite happy with getting the overview data every week from his accountant.

How Should the Data Be Presented to Him?

The simplest technique to finding the right format is to ask the stakeholders for their old reports. Start with what they know and what they're familiar with, as long as it's simple to do so.

You are probably thinking to yourself. "Hold up, surely it's better to start with something new, shiny and streamlined?"

Here is the trick though: the aim of this step is not to give them something shiny, but rather you are trying to build up trust with the client.

Let me illustrate this with a short story.

A few years ago, I heard from a designer who worked exclusively with Adobe Photoshop. He had been invited to a market research opportunity where he was given the chance to review a new rival to Photoshop. They spent a long time presenting it and showing off the product and all the features including its unique new navigation that they were extremely proud of. At the end of the session, they asked the group what improvements they would like to see. This designer said he put up his hand and said:

"Can't you just make the navigation the same as Photoshop?"

He was met with silence. To do what he asked would have delayed the product, and also meant dropping the fancy navigation feature that they were so proud of. They didn't listen to him and his primary need for continuity. The designer wanted to be able to deliver work without going through lots of additional learning.

Needless to say, the other tool isn't around anymore. Adobe's tools are still the most popular software for designers. I wonder what would have happened if that other company had listened to their target audience?

Much in the same way if you want your users to build up trust with your tools, then the simplest way to start is to replicate what they know. One of the key features that Data Studio has is that it's super quick to prototype up new reports. Once people recognize their old friendly reports sitting on their screens, you will have their attention and can start to iterate their data reports.

When we got a copy of Hermann's reporting, it looked like a big table with a wall of numbers. It was a list of products, with the numbers sold next to each item. Next to each item he could see a percentage sign that says whether sales have gone up or down from the previous week.

How Data Studio Can Help Hermann

After reviewing Hermann's answers to the four key questions, it's clear to you where Hermann needs help. He needs

1. **Timely information** – He's not getting timely information to make decisions. He needs to wait for the end of the week when his accountant sends the data back.

2. **Removing duplicated efforts** – Currently Hermann has to type his sales data into a spreadsheet and then email it to the accountant. The accountant will enter the sales into an accounting program/spreadsheet and then send back the sales reports of what's doing well. Hermann will then need to review and analyze the results.

If you can show Hermann how to use Data Studio to take his sales data and automatically process it and generate a report from the raw spreadsheet data, then you could help him to make decisions faster. He also wouldn't need to pay the accountant to enter the data and return the data back to him. Instead he can focus on driving the improvements can come from the data.

You also suspect that since Hermann is so focused on his products, he will be able to ask much better questions of his data if he is able to take control of it himself. This might be the largest win for him in the long run.

Hermann forwards you the last few reports that he received from his accountant so you can see what reporting has changed. The good news is that we can definitely improve on it! Let's crack on with the work.

Building a Report for Hermann's Cake and Coffee Shop

Step 1: Upload the Data into a Google Spreadsheet

The first step is to get Hermann's data from a Google Sheet.

For this exercise, you'll need to download a copy of Hermann's Cake and Coffee Shop sales data (DS – Coffee Shop.xlsx) from the source code for this book. Go to `www.apress.com/9781484251553` and click the download source code button.

Take your data spreadsheet and upload it to Google Drive.

Open up Google Sheets by visiting `https://docs.google.com/spreadsheets/u/0/?tgif=d`.

Create a new sheet and go to File ➤ Import ➤ Upload. Browse for the DS – Coffee Shop.xlsx file and upload it.

Make sure the spreadsheet is named "DS - Coffee Shop." If you need to fix it, double-click the file name at the top left of the sheet and change it. Hit enter or click elsewhere when you're done.

Step 2: Clean the Data

Next, we need to do clean the data so it's good to be consumed.

The aim of the cleaning process is to

- Get your data into a table format.

- Ensure that each column contains only one set of data.

- Check none of the fields have blank entries.

- Make sure that all the fields have the data in the correct consistent format. For example, you need to make sure that dates are in the format dd/mm/yyyy.

One of my favorite data team leaders used to repeat the same mantra over and over whenever developers would try to cut corners with data:

Rubbish in…Rubbish out

You should think of your Data Studio reports as being as clean as one of those fancy filtered water dispensers that you see in offices. Having water come out of them that is not clean would immediately result in them being taken out of service and an engineer being called.

This is much in the same way your Data Studio reports should be filtered and clean for your users to consume. Any data source that is infected with bad data should result in them calling in an engineer (you!) to investigate and fix the issue.

■ **Tip** Whenever you make a report, think about how you can let your users alert you if there is an issue with the data. I always put my contact details in the same place for all my reports so I am able to be contacted if people have questions. We'll talk about this in a later chapter.

Where Should Cleaning Be Done?

I know a lot of people who say that it's fine to do the cleaning of the data in the reporting tool and not in the source. I generally consider this a massive anti-pattern (a bad practice that you end up regretting).

The rule of thumb is to fix the data in source but tweak the data in Data Studio.

Data Studio has a lot of powerful features that can do some of the cleaning for you, but I tend to use these for more minor fixes or to make the data easier to consume. I recommend that people do as much of the cleaning as they can in the original source of the data as

- It keeps things simple as you only have to look in one place for any fixes to your data.

- It's less work to have to replicate your cleaning over multiple reports.

Some of the tweaks that we do in Data Studio over time, if they are clear that they are going to be there to stay, I might roll that change back into the source data just to give Data Studio that extra kick of speed.

How Is Data Cleaning Done in Professional Environments?

When you are working on doing your data cleaning in an office, you may do this work by simply modifying a spreadsheet, or in some businesses, the data teams will use a scripting language called SQL to manipulate databases with millions of lines on it.

Believe it or not, the principles behind both small spreadsheets and large databases are pretty much the same. Since SQL and databases are such massive subjects themselves (and can span volumes of their own books), we will limit ourselves hereto just doing the cleaning in the spreadsheet we created from Hermann's weekly sales data. Luckily, it's already pretty clean, so we can continue on.

Step 3: Create a Data Studio Report

Open up Data Studio and create a new blank report. Title it "DS - Coffee Shop." Refer to Chapter 1 if you need a refresher on how to create a new report.

Step 4: Create a Data Connection to the Google Sheet

The wonderful thing about data is that it's always changing. Unlike in the previous chapter, where we created a static connection to a file, in this chapter, we are building a living, breathing connection to a live Google Sheets document. When the Google Sheets data changes, then we will see our Data Studio content change too. To create a data connection to the Google Sheet, do the following:

1. Click "Add data."

2. Locate "Google Sheets" in the Google Connectors list and click it. You will need to authorize it if it's your first time connecting a Google Sheet to Data Studio.

3. Locate Hermann's sales report: "DS - Coffee Shop."

4. Select the file and select "Menu" (the first tab in the Google Sheet).

5. Click the blue connect button in the top right-hand corners, then click Add, then Add to Report.

Step 5: Create a Raw Table View of the Data

Whenever I start looking at the data that I need, I find that the easiest way to understand it is to use table control to investigate the raw data. This gives an overview of the data before we zoom down into the detail to get the information we need. It's also handy for doing some initial quality checking and double checking that the data is there and it's in the format we expect.

To create a raw table view of the data:

1. Click "Add a chart" and select the first table icon on the top row as shown in Figure 2-2.

2. Click the rectangle where you would like to place the table and you will see the data from your Google Sheet appear.

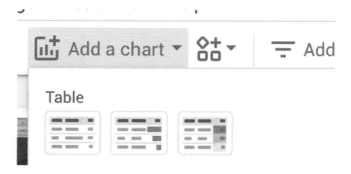

Figure 2-2. Adding a raw table to your Data Studio report

Now that we have the data successfully coming through from Google Sheets into our table, we can start to try and make the data useful to our client.

Helping Users to Ask Questions of the Data

Based on the data, we are going to try and help Hermann to understand his sales data in a bit more detail. A good starting point is to do some exploratory data analysis to understand the shape of the data. Some of the simpler questions we can answer using the data are

- What products are responsible for the majority of his sales?

- What types of flavors are customers buying the most of?

- How do the sales of these products compare vs. last year?

A raw table of data doesn't really give us a clear indication of what's going on with the sales. It just displayed to us a wall of text and numbers, which is difficult for our brains to process and interpret. We want the report to do the heavy processing for us to help our users understand the data. We can do this with by transforming the raw table into something more visual, such as a pie chart. By adding a pie chart to the Data Studio report, we are gifting the user an easier way to interpret the data. Humans are visual creatures and we relate to the world around us by using our senses to understand the world of us. The world of data is much the same; by adding visual features to the report that guide the user as to what the information is saying, it lets them focus on what is most important for them.

Add a Pie Chart

To add a pie chart to show the breakdown of sales:

1. Click "Add a chart" and select the first Pie Chart control (Figure 2-3).

Figure 2-3. Adding a pie chart

2. Click where you want to place the chart.

   When we first add the pie chart, Data Studio defaults to showing the first field in the data set which is "Product Name." It also defaults to showing us the data from this year. As you can see from Figure 2-4:

   - Our top selling product is "Banana Loaf," closely followed by the delicious Lavender Cupcake. Good old traditional Chocolate Cake is sitting in at third most popular.

   - Our top four products account for roughly 50% of our total sales.

   - Each one of our top four products is selling roughly the same proportion of units.

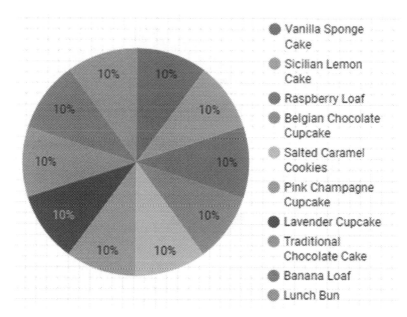

Figure 2-4. Pie chart according to product name

Step 6: Duplicate the Pie Chart and Change the Dimension to "Product Type"

Let's take a different view of the data to get a different perspective by duplicating the pie chart and changing the dimension to product type (Figure 2-5). To do this:

1. Right-click the pie chart and select Duplicate.

2. Drag the pie chart immediately to the right of the original pie chart.

3. Select the new pie chart and click "Product Name" in the right Panel under "Dimension" and select "Product Type."

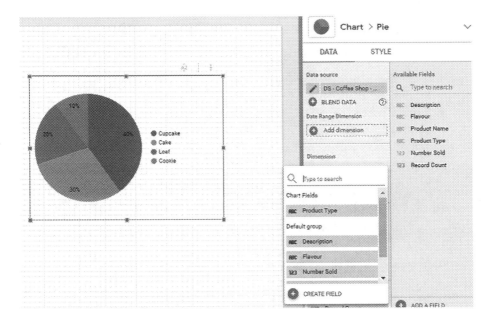

Figure 2-5. Pie chart according to Product Type

When we selected the dimension called "Product Type," we told Data Studio to reprocess the data in our table, and instead of grouping it by product name, we should instead group it by product type. By changing this "dimension" that it uses, it changes the lens that we use to look at the data and shows it from a different perspective. Dimensions (and metrics) are the fundamental building blocks for data reporting, and we will be exploring these further in the next few chapters in the book.

For now, let's just admire how the small change to our pie chart has helped us to understand the sales data even more. When we view the sales data through the lens of "product type," we can see

- Nearly 50% of our sales come surprisingly from cupcakes.

- A third of our sales come from cakes.

Step 7: Use Scorecard Control to Show Sales

With all reports, it's important to recognize that there is a hierarchy of importance. Think back to your report cards that you received from your teacher at school. The most important information that you wanted to know was what your grade was for each subject. How would you have felt if the grade was hidden away at the bottom of several paragraphs of comments about how hard you worked? You would have been very frustrated and confused about why they had hidden away such important information. Much in the same way, we need to use scorecards to highlight the most important information to readers of our report.

Typically use scorecards to show off what we typically call "KPIs," or key performance indicators.

The best and most concise definition of KPI I have seen comes from the UK's Company Act where they define KPIs as "factors by reference to which the development, performance or position of the business of the company can be measured effectively."

To add a scorecard that shows the number sold to your report:

1. Click the "Add a chart" button in the main toolbar.

2. Select the first icon below the word "Scorecard."

3. Click and drag a box to indicate where you want the scorecard placed (ideally in the top left of the report).

4. By default, it'll show the record count for the product names, so you'll need to change the dimension to number sold.

You will now see the scorecard with the "Number Sold" label (Figure 2-6).

Figure 2-6. Scorecard for number sold

For Hermann, his KPI is going to be how much product he sells so this is something to show immediately in the report. We use the scorecard control to show the number of products sold at the top of the report.

■ **Tip** When it comes to putting scorecards onto your reports, try to use as few as possible. Too many scorecards can confuse the user. If you find you are putting more than five scorecards on a report, it's time to think whether you should use an alternative presentation like a table or bar chart.

Step 8: Add a Date Comparison to the Scorecard

Now that we have used pie charts to show how big the sales of each product are, we can move on to the next stage of our analysis. Data is always changing. What is true for today is not always guaranteed to be true tomorrow.

I don't set trends. I just find out what they are and exploit them.

—Dick Clark, American radio personality

People's fashion, tastes, and trends change regularly, and it's up to businesses to use data to respond to these changes. If you are able to follow the changes in your customers' tastes or even seek out new customers' tastes to exploit, then you are winning.

We are going to help Hermann start on his journey of understanding what trends are affecting his business and give him an idea of what has changed over time with his sales. We hope that by doing this we might be able to identify some opportunities that he hasn't thought of yet.

We know from inspecting our Google Sheets data that we have a field for "date." Using "date/time" data is really powerful as we can explore what changes have happened between last year and this year. We are going to start by showing him how much sales have changed over time. We are now going to choose the date period that we are going to compare against. Select the "comparison data range" button. We can choose to compare against

- None
- Previous period
- Previous year
- Fixed/advanced

The "previous period" option is perhaps one of the most powerful features in Data Studio. It will take the current time period that you are looking at and compare it against the previous equivalent time period. You can thus use this control to compare week on week, month on month, or even year on year to see how your sales are changing.

We want to keep things simple for now so we are going to select "Last Year" and press "Apply" to complete our "Year on Year Scorecard."

After a second or two, you will see below your scorecard a green arrow appear next to "47%" (Figure 2-7). This shows that the number of products sold has gone up 47% since last year which is an amazing result for Hermann.

Figure 2-7. Scorecard showing comparison

Now that you know how to do basic comparisons with date ranges on scorecards, the good news is that you can do the same thing for tables too. Why don't you try do the comparison year on year for the table of data you have and see which of Hermann's products are responsible for driving his improvement in sales since last year.

Back to the Books

One of the most valuable things about trying to use data to solve a "real" world problem as we did in this chapter is that you begin to realize that there are gaps in your knowledge and there are new things that you need to learn.

From your first experiences using the Data Studio tool, you probably noticed there are so many new concepts and terms being used all over the place. Don't worry, you'll get the hang of them as we go through the book.

▓ **Tip**   Data Studio's documentation team also maintains a handy glossary of terms (`https://support.google.com/datastudio/topic/6302375`).

Recap

Doing something simple like understanding more about your data is vital to every business. In this chapter, we took a simple Google Sheet of data, and instead of leaving it as a static data set, we used Data Studio to bring the data alive. We could identify

- What products were selling the most for the client
- What product types were most popular
- How much sales had changed over time

We tried out some simple controls such as tables, scorecards, and pie charts and even started doing some basic comparisons of this year vs. last year.

In the next chapters, we will expand on this, but hopefully you now have a good feel for using the Data Studio interface to explore your data.

Metrics and Dimensions

Helpful Data Visualizations

You know you live in the world of Web Insights when you realize that every piece of data you look at drives action….

—Avinash Kaushik, Digital Marketing Evangelist, Google[1]

In the previous chapter, we needed to use dimensions and metrics to make our reports. Let's continue trying to understand what these are.

Just about anything to do with data can be boiled down to the metrics and dimensions.

Metrics

Data Studio's Help center glossary defines metrics in a really overly complicated way:

---

[1] https://www.kaushik.net/avinash/traditional-web-analytics-is-dead/

© Grant Kemp, Gerry White 2021
G. Kemp and G. White, *Google Data Studio for Beginners*,
https://doi.org/10.1007/978-1-4842-5156-0_3

A specific aggregation applied to a set of values. Metrics are aggregations that come from the underlying data set, or are the result of implicitly or explicitly applying an aggregation function, such as COUNT(), SUM(), or AVG(). The metric itself has no defined set of values, so you can't group by it, as you can with a dimension.[2]

Let's try to simplify this definition. Metrics are a way of apportioning a value to the size of an item. A metric can be succinctly boiled down to

How many of a certain thing you have.

Consider Hermann from Chapter 1. As the owner of a modern cake shop, he is interested in counting how many cakes are sold in a particular day or how many ingredients they have used up.

Other examples of metrics include

- The temperature on a thermometer
- The number of points awarded in a game show
- Number of people in a medical study
- Website users, that is, how many people arrived on your website in a given data period
- Sales revenue from selling your products online
- Number of errors shown to users in a day
- The average amount of rain in a region

Aggregating Metrics

Metrics can also be much more sophisticated. We could be interested in understanding how the numbers change over time or how they compare between time periods.

The owner of the cake shop could simply measure the number of cakes sold, but he could also be interested in finding out how many of the cakes were sold in a given day.

1. Whether sales were increased or decreased compared to a certain period of time, and are they growing slowly or are they declining rapidly

2. How the number of sales compared to last week, last month

---

[2]https://support.google.com/datastudio/answer/6380885?hl=en

When you analyze a "collection" of metrics, this is known as aggregating them or bringing them together. Aggregations become themselves new metrics. Aggregating allows us to compare, analyze, and trend data to find insight.

Note that it is possible for a metric to be one raw number or it can be an aggregation. An aggregation can become a new metric, and even while using other metrics within its formula, it becomes itself a metric.

We will come back to aggregations in more detail later on, but the Data Studio glossary definition of metric mentions three:

- **Count() or Counting metrics** – An example of this would be how many cakes were sold in a given day.

- **Sum() or Sum of metrics** – This could be the sum of all the revenue earned from selling cakes in the cake shop through cash and credit card sales.

- **Avg() or Average of metrics** – This could be the average number of bakery staff (bakers, cashiers, etc.) working in a week.

Dimensions

Unfortunately, a metric or aggregation doesn't give you the full picture when it comes to understanding your data. You need to have more information in order to work out whether the data you have is useful.

With metrics we are able to really understand the quantity side, but how do we understand how to put a value on quality? This is where dimensions come to our rescue.

The Data Studio glossary defines dimensions as being:

A set of values by which you can group your data. Dimensions are categories of information. The values contained in those categories are typically names, descriptions or other characteristics of that data.[3]

This is a pretty simple explanation, but I think we can do a little better at making it easy to understand:

Dimensions are the names, descriptions or other characteristics of the things you are counting.

---

[3]https://support.google.com/datastudio/answer/6303401?hl=en

Consider cavemen foragers. Just knowing that they have collected 100 berries in a day isn't the difference between life and death. Knowing that they have collected 50 delicious berries and 50 poisonous berries is the difference between life and death. This crucial extra "dimension" to the data immediately makes the information far more valuable and useful. It will help them in the future alter the bushes they chose to collect berries from and give them a "healthier" outlook for the future.

Unlike cavemen, for most modern people, their data won't be the difference between literal life and death but rather will be whether their business succeeds or fails in its endeavors. We all want to win in our work life, so making sure that we have the right dimensions to understand good from bad will be a crucial part of our job.

Recall the examples of metrics we covered in the previous section, and let's add dimensions to some of them. Knowing the dimensions changes the information that you are receiving from a metric. The dimensions are bolded in the following list:

- The number of goals or points scored at a sports match **by each team**

- Number of people in a medical study as well as understanding their **age, height, or fitness level**

- Website users separated **by country of origin, type of device they use, or whether they buy or not**

- Sales revenue from selling your products online, broken down into **sales per country or sales per category**

- Number of errors shown to users in a day broken out **by error type and how severe the errors are at impacting customers**

If you are ever struggling to find the dimensions of your data, just describe it to yourself. If you hear magic phrases like "broken out by" or "per," you will know that these elements are dividing up the data and can, in most cases, give you better idea about how dimensions are influencing the metrics you are counting.

Let's think about Hermann's cake shop. What information can we think about when it comes to our cakes that can help understand a bit more about what is selling well?

We know that we are selling cakes, but there are other dimensions that we can break out from the data to give us a little more context. Just knowing that we are selling 100 cakes doesn't really help our team to know what types of cakes that they should produce in the future.

You can usually find out what are the most important characteristics of the cakes you sell by looking at the product you are selling or asking the business or its customers to give you some information about what they think is the "deciding factor" that influences the sales of the product. Just asking the baker what the differences are between the products is a great start.

- What size are the cakes? Are they teacakes, cupcakes, or multitiered wedding cakes?

- What flavors are the cakes? Chocolate, Vanilla, Carrot, or Banana flavor?

- What theme is the decoration for the cake? Is it a plain cake with just frosting on top, or is exquisitely decorated for a couple's forthcoming wedding?

As you can see from the bakery example, there are lots of deciding factors that can be described in relation to the sales we are making. Understanding that certain flavors are more popular is a vital piece of knowledge, as it means that the baker can showcase these most popular flavors.

One final thing to keep in mind: most modern businesses analyze data that is "multidimensional" – meaning it doesn't just rely on a single dimension to explain the data. Thinking about our cakes example, the bakery will sell a multitude of cupcake flavors, but with each of these flavors, we might have a multitude of different flavors of buttercream icing. We might also have a multitude of different themes, and each of these themes may sell differently depending on the day of the week or season. For example, sugar-free cakes will likely sell well to certain populations.

This is where being a good data developer or data visualization engineer comes in. Dimensions add context to a metric and bring understanding to the data. You have the ability to "try on" a set of dimensions on a data set to see if it provides more insight for the end user. You can also see if the data has a strong correlation with the business goals that you are trying to achieve.

Recap

In this chapter, we talked about how metrics by themselves or aggregated can be valuable, especially when given context by considering dimensions as well.

An Online Survey

A Case Study Profile Analysis

> *It is, in fact, nothing short of a miracle that the modern methods of instruction have not yet entirely strangled the holy curiosity of inquiry.*
>
> —Albert Einstein, theoretical physicist

Now that we've covered some basics in the previous chapters, let's see how we can use Google Data Studio for another real-world situation: analyzing the data resulting from an online survey to gather customer information for a client. This chapter is based on a real-life project that we did for a popular travel company. For a client to get an interactive analysis of their customer data without expensive visualization tools is not only cost-efficient but also incredibly useful. It gave our client insight that they could use to develop their content and marketing strategy to better target the customers.

Finding Out What Travel Company Needs

First, we needed to find out more about our client's needs. As a popular travel company, they wanted us to take a look at who their customers were.

© Grant Kemp, Gerry White 2021
G. Kemp and G. White, *Google Data Studio for Beginners*,
https://doi.org/10.1007/978-1-4842-5156-0_4

Gathering the Data

The next step was to gather the data. We sent about ten thousand emails to our client's customers and asked them to fill out a survey using SurveyMonkey.

The survey asked questions about the customers themselves:

- Gender
- Age range
- Salary range
- Marital status
- Kids (and grandkids)
- Job title

And about their view of the website/service:

- How often they holidayed
- Reasons for going on a holiday
- Did they like aspects of the website
- Did they like aspects of the service

We also had free text questions, a text box allowing them to type information such as if they thought something was missing from the website, allowing more open responses to some key questions.

Once all submissions were received (about 5,000 surveys), we're ready to prepare the data for the report.

Preparing the Data

The next step was to prepare the data before we could feed it into Data Studio. SurveyMonkey is able to export the results into an xls file, which is a format we could easily import into Google Sheets. After uploading the results to Google Sheets, we started cleaning up the data.

As discussed in Chapter 2, clean up entails ensuring that nothing in the data set is invalid. For instance, despite us using the company's mailing list, presumably of people who wish to be contact about the service, we still received "junk" responses from people who did not want to be bothered. We also had to remove anything that is considered a null, or blank, response. We also needed to make sure that all columns have unique names. Don't worry – every data set has issues.

In this case, we were dealing with a relatively small data set of about 5,000 surveys, and we can pull it directly from Google Sheets into Data Studio. If we had a larger data set, we could upload it into Google BigQuery first and pull it into Data Studio in a similar way.

After we cleaned up the data, we were then ready to pull it into Data Studio. Make sure you have access to the Google Sheet in the same account and the file name is appropriate.

In Data Studio, start by creating a report. It will then ask you to add a data source to the report. Simply click the Google Sheets option (Figure 4-1).

Figure 4-1. The button to click for Google Sheets

Designing the Report

Next, we need to design the report to make it useful and aesthetically pleasing. We can do this by structuring the presentation of data and making elements consistent throughout.

It's always nice to have elements that are consistent through the report. This promotes a sense of cohesiveness, a connection to the company's branding, and it is aesthetically pleasing.

Getting things looking consistent means that your clients will find the report easier to read.

Header Bar

One thing you can do is add a header bar that has the company's logo. You can do this with a plain text header as well, but a logo looks nice if you have it.

To upload a graphic, click the Image icon in the toolbar and select "upload from computer" or "by URL," whichever is applicable for you. Then, place the logo somewhere at the top of the report page. Right-click the logo and select "Make report level." This means that the logo will appear at the top of every page if your report has more than one page.

Color Theme

Another thing you can do is set a color theme for the report so it is consistent throughout. If the company you are producing the report for has a color scheme, you can incorporate it into your report by extracting the color scheme from an image, which can be the company's logo or even a screenshot of their website.

On the toolbar click Theme and Layout, and on the right pane click "Extract theme from image" (Figure 4-2). You can either upload the image from computer or by URL.

Figure 4-2. Screenshot of button to extract an image palette

You can then select from three generated themes. If you like one, click it and press apply. If you find the output of the extraction doesn't have enough color diversity or contrast, you can manually edit the color theme to make your report better fit your company or clients style. Click the customize button in the Theme and Layout pane.

Footer Bar

Also consider creating a footer at the bottom of the report with information on how to contact you, or information here on how to get elements fixed up, if you are an agency or in house team a link to your contact details if someone wants something fixed.

You can use the text tool to type in this information and place it at the bottom of the page.

Detailed to Granular

Plan to present the data with an overview at the top, then work your way down to more detailed and granular data. It is often easier to start with a longer page, then trim it down from there.

I would also recommend using a rule of thirds or 50/50 for content – meaning that you make tables and charts either one-third, two-thirds, or 50/50 – you can draw a line two-thirds across the report and halfway so you can use these as guidelines for layout. You can use the Line tool in the toolbar for this purpose.

Getting Data onto the Page

Next, we need to add data to the report. As we mentioned earlier, it is a good idea to start with overview data before drilling down into the details. A good way to do this is to start with a pie chart.

Although pie charts are often considered to be bad for users as it is hard to interpret exact values, you can use them (or a doughnut chart) at the top for key groups to be used as filters. Almost any chart can be used with a filter by setting the Interactions – Apply Filter to be enabled. This will mean that if you click a segment within the doughnut or pie, it will apply this to the data source.

Generally, we will apply this to all the other charts and tables – allowing them all to be able to filter the report as people click into rows or slices of data (see Figure 4-3).

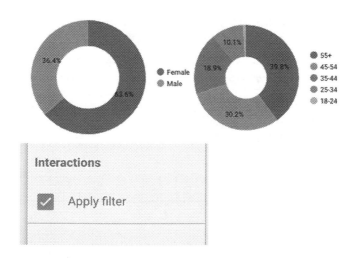

Figure 4-3. Apply Filter button

We now have a very visual way of seeing the data within the report and being able to quickly drill downward across all tables and information from the same data source. If we have multiple data sources, it will normally only apply to elements from the same source (you can also use groups of charts if you do not want filters to apply to the entire report).

A list of this information alone is potentially useful to a marketing manager, but they probably already know most of this, and so to make it more useful, we combine dimensions together; it gives us a little more insight. We can do this using pivot tables or more graphically using stack charts. For example, if we use a stacked bar chart, we can combine our data on gender, marital status, and if they have kids. See Figure 4-4.

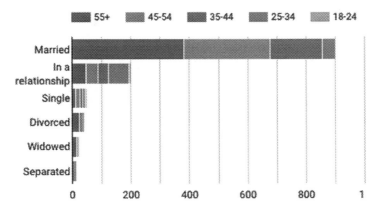

Figure 4-4. Stacked bar chart

At this point now we can see that most of the respondents in this form are married and most are over 55 years old. This prompts us to consider: *Is any bias in the data? Are these people more likely to fill in a survey?*

A pivot table can also give us greater insights, especially if we use a heatmap. This combined with the questions that marketers can use to make decisions.

The pivot table heatmap (Figure 4-5) combines purchase frequency with demographic information. According to this, the marketing should target people who are more mature for frequent loyal customers.

Age / Count					
Purchase Frequency	55+	45-54	35-44	25-34	18-24
5	8.93%	7.79%	4.51%	0.9%	0.08%
4	4.51%	3.44%	2.3%	1.07%	0.08%
3	5.66%	5.08%	2.62%	1.15%	0.16%
2	6.56%	5.25%	3.44%	2.38%	0.25%
1	7.7%	4.84%	3.93%	2.87%	0.08%
0	6.48%	3.85%	2.13%	1.72%	0.25%

How often have you purchased in the past five years?

Figure 4-5. Example of a pivot table showing purchase frequency by age

You can combine this table with other relevant visualizations (Figure 4-6) and watch how when interacting with one it impacts another. These relationships with data can be quite hypnotic, and the easiest visualization for creating these interactions is pie charts. However, as mentioned before, we should try to avoid pie charts as humans are not very good at reading and comparing angles. Many experts suggest using either a treemap or a bar chart instead for simple filters. Despite that, pie charts are still a decent alternative to using a drop-down box.

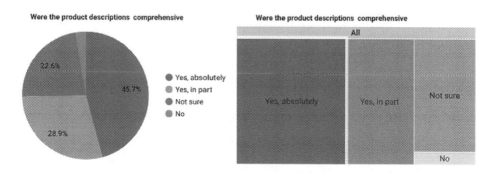

Figure 4-6. Pie chart and Treemap demonstrating how to show data that can be used as a filter

Either type can be made to apply a filter, which means that they can be used alongside the more detailed information that is driving decisions. For instance, they can be used to understand the impact of a particular feature on a website.

Free Text Analysis

There were a number of free text boxes in the survey that we can look at in several different ways. If there are enough entries, it's possible that there will be some overlap or essentially the same response. It can be a little challenging, but with a little analysis we can take the data a little further.

In the survey, we asked if there was anything missing in the product description. Figure 4-7 shows the top results.

Was there anything missing?	Count ▾
null	668
Yes	47
yes	25
No	17
Nothing missing	5
Fine	4
N/A	3
All good	3
n/a	3
good	3
N/a	3
It was fine	3
Was fine	3
Nothing missing.	3

Figure 4-7. Table showing free text results for the question "Was there anything missing from the product description?"

Ideally we would like to consolidate similar results and remove the "null," or replace it with no response.

- Lowercase and Trim all responses
- Case statement to consolidate common patterns

Ultimately it would be good to do smart extraction using machine learning, which can be done cheaply and increasingly easily, but that's the subject for a separate book. For now, we can work with the tools in Data Studio.

"Null" can be easily replaced by "no data" in the formatting (missing data), and while we are there, we can improve the text handling, changing it to "Wrap Text" for longer answers. See Figure 4-8.

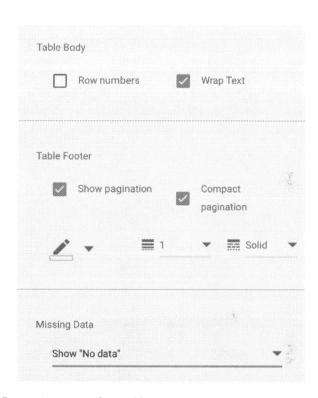

Figure 4-8. Formatting options for a table

We can also include additional classifications for common words. There are some word cloud visualizations within the community selectors, but none of these quite meet our needs. So the first part of this is almost by-hand analysis; when we spot common patterns, we can use the case statement – if there are particular themes we want to bring out, for instance – some mentioned images, some mentioned pools, and some mentioned kitchens.

In Figure 4-9, we can see how we can classify the text if it includes a specific word, which allows us to filter the report by people who entered a particular keyword such as in Figure 4-10.

Field Name

Missing Themes

Field ID

calc_eddco3tl7b

Formula ⑦

```
1  CASE
2  WHEN REGEXP_MATCH( Anything Missing , ".*images.*") THEN "Images"
3  WHEN REGEXP_MATCH( Anything Missing , ".*photos.*") THEN "Images"
4  WHEN REGEXP_MATCH( Anything Missing , ".*pool.*") THEN "pool"
5  WHEN REGEXP_MATCH( Anything Missing , ".*kitchen.*") THEN "kitchen"
6  ELSE "Unmatched" END
```

Figure 4-9. Custom field with case statement

One minor issue is that using this method, they can only fit into one theme, and so if it mentioned photos or images, then it won't also be counted as being in pool; as such, the order/prioritization of the case rules is important. Besides, this does give us the ability to create a simple drop-down so you can spot and analyze common themes, and of course this drop-down will also apply to all other data in the report. See Figure 4-10.

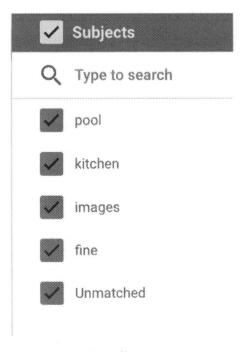

Figure 4-10. Classified text in a drop-down filter

Final Result

So now we have an interactive report based on survey data that allows a marketer to drill into this and develop insights of their own, the best thing – it is quite easy to tweak and improve. See Figure 4-11.

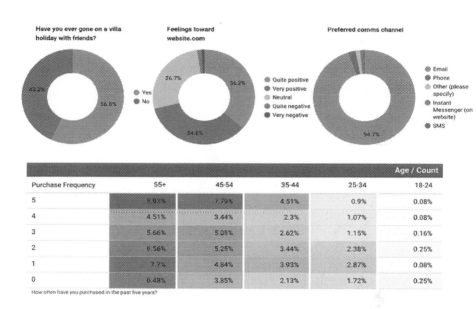

Figure 4-11. Outputting the data with heatmap pivot tables and doughnut charts

Recap

Google Data Studio isn't just a front end for your analytics data. It can be a presentation layer for survey data, much like a tool like Excel can be, but makes the data online and accessible for everyone.

Data can be improved and cleaned within the tool using simple string functions, and additional transformations can be applied to create custom fields based on free text or by consolidating some dimensions. Ultimately Google Data Studio is a powerful way of taking data sets and presenting them in easy-to-understand ways. For some this is a far better front-end to complex back-end tools such as Google Sheets, but care must be taken to make sure the data is in the right format to start.

Marketing Data Visualization

Would you tell me, please, which way I ought to go from here?

That depends a good deal on where you want to get to, said the Cat.

I don't much care where—said Alice.

Then it doesn't matter which way you go, said the Cat.

—Alice's Adventures in Wonderland, Lewis Carroll

In this chapter, we are going to zoom in on some of the best and most potentially transformative source of data in the world, namely, website analytics (also known as on-site analytics), and learn how to present the date in a report.

The Value of Website Analytics

The majority of businesses are data factories, spilling out conveyor belts of data about what's going on with the business. Like a real factory, these conveyor belts never stop unless someone breaks something or the business stops trading. It's the job of the data visualization engineer to turn these conveyor belts of data into insights that can be used by business to drive

G. Kemp and G. White, *Google Data Studio for Beginners*,
https://doi.org/10.1007/978-1-4842-5156-0_5

value. Molding products out of these multitude of data sources will be one of the most valuable contributions that your new Data Studio skills will be able to process, mix up, and merge, which will drive a lot of insight and impact.

Webstores are now considered significant contributors to a company's business, so looking at the web analytics for a webstore is valuable because you can then understand how and why the website works and how it can be made better to drive even more business.

Website analytics offer so many wonderful opportunities to fundamentally change businesses performance, and yet so many businesses make the same error of simply downplaying it or worse ignoring it all together.

For me the simple definition of what on-site web analytics is:

What's going on with my website?

Entire books have been written on the subject of analytics, which go into massive amounts of detail. I am not going to regurgitate the same information here but rather direct you to a source that will give you a high-level summary.

Check out the Wikipedia web analytics page which is in my opinion one of the best worded definitions out there: https://en.wikipedia.org/wiki/Web_analytics#Web_analytics_technologies.

I recommend that you skim through it to try and give yourself some context about of analytics and in particular have a read of the section on the more common "On-site web analytics."

Visualizing the Size of Your Audience

Web analytics can affect a large number of potential customers that it's difficult to understand just how powerful it can be. 3,000 people can look very similar to 30,000 people if you are only looking at numbers.

Here's a real-life example. After spending time looking at the data in the web analytics for a client, I recommended that we add a "Refer a friend" program to a furniture retailer. We don't have to go into why, but let's just say the data pointed me in that direction. Since they literally had millions of visitors coming to their site every month, the referral program was able to scale really quickly. It was only a month or two later where we found that they had acquired a lot of new customers really quickly. When I showed them the number of new customers in a slide deck, I remember getting a lukewarm reaction as it was still considered only a small channel for the business.

When I presented the numbers a few weeks later, I tried to make the report more engaging. Instead of reporting the actual number of new customers, I showed them a picture of a particular football stadium and said we could fill one of them twice over with the new customers that have come in from the "Refer a friend" program. The reaction was to applaud! This was the same data, and same results presented to two similar teams, but presented in different ways.

Google Analytics

Google Analytics has become the most popular choice for web analytics since it was first created over 10 years ago. It's quite difficult to find a website where there isn't some portion of the site tracked using its code. Arguably one of the key drawbacks about Google Analytics and one of the main reasons why Data Studio was born was so as to create a simple and engaging platform for dashboards. Figure 5-1 shows the old Google Analytics dashboard.

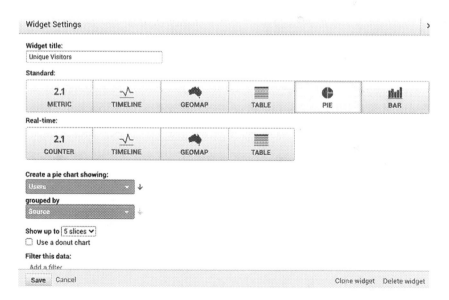

Figure 5-1. The old Google Analytics dashboard

Google Analytics has had built-in features for dashboarding for many years; however, in most countries, this has never had widespread usage because

- It required users to log in to Google Analytics.
- Google Analytics dashboards are complicated to share and change.
- You are limited in the amount of data controls that you can use.

Let's get started visualizing with Google Analytics.

Starting Out Visualizing Google Analytics

The first thing to do is get data to analyze. For the sake of learning, we are going to use some practice data, and so we can get started quickly, we are going to use templates.

Getting Access to Practice Marketing Data

Before we get started with some data wrangling, we will need to get hold of some data that we can practice on. The Google Analytics team has provided several data sets that can be used in Data Studio to help us all to learn.

The data set that we are going to use for the next exercise is the Google Merchandise Store data, which is made up of all the visitors and customers who visit https://shop.googlemerchandisestore.com/. See Figure 5-2. The great thing about this data is that it effectively allows us to learn what real website data looks like without having to invest thousands in building out our own site. One point to note is that there are usually some oddities that occasionally crop up in the data which don't normally make sense; however, I am pretty sure that this is down to the data being created for the purpose of demonstrating certain features. Generally speaking, the data is perfect for most learning purposes; however, be wary of comparing it to your own. Also don't worry about the potential risk of breaking the account. There is nothing you can do wrong that could damage the data.

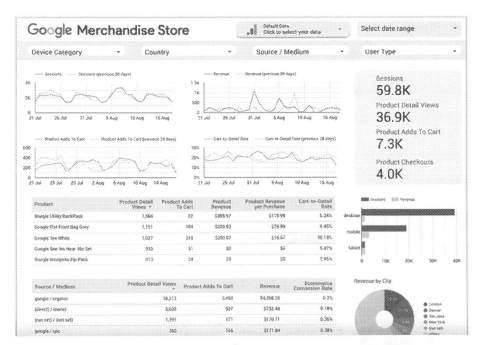

Figure 5-2. The Google Analytics Merchandise Report from the Template Gallery

To Get Access to Google Analytics Demo Data

Accessing the Demo data can be done via a link.

1.   Make sure you have registered for a Google Analytics account by visiting https://analytics.google.com and signing in or creating an account.

2.   Visit the Demo Account Help Center article (https://support.google.com/analytics/answer/6367342).

3.   Click the button that says "Access Demo Account."

4.   This will add the demo account to your Google Analytics account. You will be able to confirm this by clicking the "account picker" at the top of the page and selecting the "Demo Account" as shown in Figure 5-3.

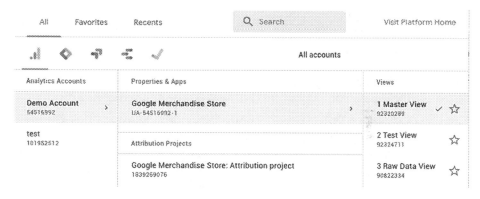

Figure 5-3. The Google Analytics Demo account

If you were using real Google Analytics data, you would do the same steps as the aforementioned except with a real Google Analytics account.

Creating a Report with a Data Studio Template

Templates are really useful when it comes to getting a jump start on building your own Data Studio reports. You can use it to learn from some of the best in the industry about the right way to do things. Let's start by learning how to build your own reports from the Gallery.

Select the Template Gallery button on the right-hand side of the screen.

When you select the Template Gallery button, you will see a full list of the various reports that have been made by the Google Analytics team (Figure 5-4).

Figure 5-4. The Template Gallery has a wide selection of templates you can use

Select Google Merchandise Store as the template on the bottom row. Then Select "Use Template" on the blue bar across the page.

Once you have selected the relevant report and chosen "Use Template," you are given the option to either use an existing source or create your own.

Since not all of you will have access to an analytics account with live data, Google have luckily managed to share the data from their Google Merchandise Store – which is a great resource to try out when you are learning.

Next, click "Select a Data Source" under "New Data Source." You will see a screen similar to Figure 5-5. Simply select the Google Analytics Demo data from the drop-down menu below "Data Source."

Copy this report

Select a data source(s) to be added to the new report.

Original Data Source	New Data Source
(Unknown)	Select a datasource... ▾

Note that **report editors** can create charts using the new data sources and can add dimensions and metrics not currently included in the report.

Cancel   Copy Report

Figure 5-5. The copy report options

Select Copy Report and double-click the report name and label it as "GA Learning Report."

You are then good to go to start exploring the new report. I thoroughly encourage you all to click the various reports and look at the properties for each of these reports in the "Properties Panel."

You now can use any of the existing templates that Google has provided to create your own report.

Practicing with More Demo Data

The best way to learn any new skill is to copy what an expert does enough times so that it becomes second nature to you. This will let you start to build up your Data Studio skills so it become second nature to you and you can focus more on the content of the report than the making of the report.

Luckily Data Studio has two main places where you can learn how to visualize data from people who use it all the time:

- The Template Gallery (https://datastudio.google.com/ u/0/navigation/reporting)

- The Data Studio Report Gallery (https://datastudio. google.com/gallery)

Right now we are going to focus on the Template Gallery to get us up and running. For those of you who are looking to build your own reports for your websites, the Google team produces a lot of premade reports via its Template Gallery. The Template Gallery is shown in Figure 5-6.

Figure 5-6. The Template Gallery

You can usually find these listed along the top of the page when you browse to the Data Studio home page.

Data Studio is always evolving, and there are always new improvements that you can add to your reports.

Recap

In this chapter, we talked about the value of website analytics, and we tested out visualizing Google Analytics using a demo account and templates.

Blends of Data

Combining Sources to Gain Better Insight

The important thing is to not stop questioning. Curiosity has its own reason for existing.

—Albert Einstein, theoretical physicist

Often a single source of data gives you some helpful insights, and by now you'll have learned that including multiple data sources into a report allows you to present more of a complete picture. But when you blend multiple data sources into a single table, graph, or calculation, this can bring far more insight into the relationships and you can answer more of the why questions.

Some questions blends can answer:

- Is our store traffic dropping on hot and sunny days?

- Is the paid brand traffic impacting our brand organic?

- Can I see a table with organic impressions, clicks, and conversions by page?

Ideas of what can be blended are limitless, and for digital marketers, especially in SEO (search engine optimization), this will often lead you to pull data in from multiple sources such as crawler data, weather, product stock data, and more.

© Grant Kemp, Gerry White 2021
G. Kemp and G. White, *Google Data Studio for Beginners*,
https://doi.org/10.1007/978-1-4842-5156-0_6

In this chapter, we will go through the process of making some simple blends to see if we can give better insight and work through some real-world problems and solutions.

Google Data Studio Blends use something called an outer left join, which means that, for example, if you are combining two sets of data about a web page, the first set will have all the rows in it but the second will only have where the two have a match.

You can blend two or three different analytics accounts, for instance, which would allow you to see a cumulative or comparative view of data. This is very useful within a migration or when you have a different account on different subdomains. Although, it is worth noting that some sitewide metrics won't be available if users cross multiple properties.

The First Blend: Select Two and Blend Data

There are two approaches to blending data. The one that is easiest and often a way to start is to just blend two charts together. If you have two charts that you would like to combine, simply right select both (select the first, hold down shift, and select the second) and then right-click (or control click a Mac) and "Blend data" (Figure 6-1). If Google Data Studio can find a common key such as date, it will magically just work, amazing – suddenly we have two data sources over one chart.

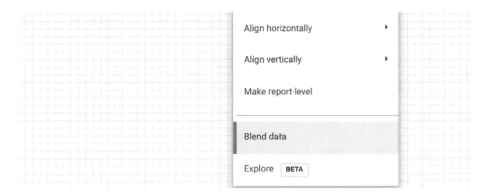

Figure 6-1. Blending data with two charts selected

If you find that one source is at a completely different scale to the other, such as if we combined traffic to the website and weather, we can assign one of the series to the right axis of a time series chart, allowing for both to be displayed at a scale more appropriate. You can access these options in the Chart ➤ Table ➤ Style pane (Figure 6-2). So simple, so powerful, and works with most time-based charts. If you've combined two charts to make a third, you can successfully delete the first two.

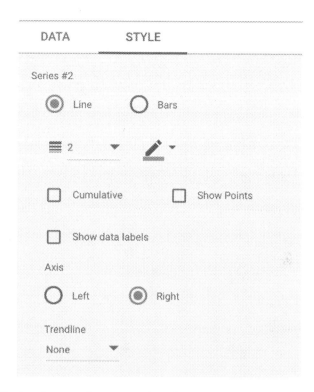

Figure 6-2. Moving the second metric to the right axis

Adding in More Data or Changing the Blends

If we combined two data sources using the first method (two charts and hitting blend data option), we might discover we wanted to add a filter, or another metric to the graph which isn't there, we can edit this blend. The blended data source is added to your list of data sources in the same way the first two are – it is typically badly named "Blended data (1)." Let us change the name of it first. If we click the pencil next to the data source (Figure 6-3), on the right side of the screen pane, you can see "data source name" and edit it (Figure 6-4).

Figure 6-3. Edit the data source name

Figure 6-4. A data blend's initial name

We can add in more metrics and dimensions providing the matchup will allow it; if the data doesn't have the right key or if the data can't be matched, this of course won't work or will give you incorrect results.

An example would be metrics that match with a landing page including conversion rate and bounce rate – but if we were using page view metrics, these wouldn't be correct.

We can edit the name of the new connection, add in dimensions and add in metrics, and we can also add in calculated metrics that use both of the original connections – for instance, we could find the ratio of clicks to sessions from Google Search Console to Google Analytics, how closely organic traffic matches up (Figure 6-5).

clicks to sessions|

Formula

1	Sessions / Url Clicks

Figure 6-5. A calculated metric in a blend

The Key to Blending

Remember that the blend needs something to key into – this can be as simple as date which will usually match if the two data sources overlap and are both similarly granular. For anything more complicated, we might need to modify the data slightly to make sure both sides match.

An example would be taking the page level data from Google Search Console (GSC) and the data from Google Analytics. We can use the Google Demo account for this example (Figure 6-6). Two things we can see quite quickly are that in Search Console it is the full URL but in Google Analytics it is just the page path that is tracked.

	Landing Page	Sessions ▾
1.	/home	32,070
2.	/google+redesign/shop+by+brand/youtube	4,403
3.	/google+redesign/apparel/mens/mens+t+shirts	2,242
4.	/store.html	2,060
5.	/google+redesign/apparel	1,840
6.	/google+redesign/new	1,446
7.	/google+redesign/apparel/mens	1,068
8.	/basket.html	926

1 - 100 / 439 ‹ ›

	Landing Page	Url Clicks ▾
1.	https://shop.googlemerchandisestore.com/Google+Redesign/Shop-by-Brand/YouTube	3,026
2.	https://shop.googlemerchandisestore.com/Google+Redesign/Apparel/Mens/Mens+T+Shirts	2,450
3.	https://shop.googlemerchandisestore.com/Google+Redesign/Apparel	1,471
4.	https://shop.googlemerchandisestore.com/store.html?vid=20180201712	818
5.	https://shop.googlemerchandisestore.com/Google+Redesign/Apparel/Mens	599
6.	https://shop.googlemerchandisestore.com/Google+Redesign/Stationery/Stickers	574
7.	https://shop.googlemerchandisestore.com/Google+Redesign/bags/backpacks/	470
8.	https://shop.googlemerchandisestore.com/store-policies/frequently-asked-questions/	355

1 - 100 / 648 ‹ ›

Figure 6-6. Tables comparing the dimensions that will be blended

We can also see that capitalization in GSC is different to that in Google Analytics, as someone has filtered it; these discrepancies make blending a little more challenging, but both are easy to overcome. Figure 6-7 shows how you would edit a field.

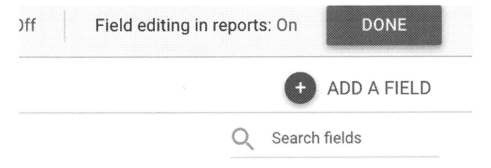

Figure 6-7. Editing a field

We can also use string functions to lowercase everything and remove the domain so both "landing pages" are the same.

Manually Creating Blends

We might want to combine a page's metrics from both Google Search Console and Google Analytics, but if we do this, we run into a couple of initial issues – first add your first data source – Google Search Console to the page – we are using the Google demo account for this.

We want to see the impressions, clicks from Google Search Console, bounce rate and conversion rate from Google Analytics in a single table.

Let's take this page:

shop.googlemerchandisestore.com/Google+Redesign/Shop+by+Brand/ YouTube

In Google Search Console, this is the landing page URL, but in Google Analytics, the landing page looks like this:/google+redesign/shop+by+brand/ youtube

All lowercase and without the domain (and subdomain), so we edit the data source (though we can't do this with the demo account, so just follow along). First click the pencil from the data source already added to your report; second, we need to add a new field (as we can't modify the source) by clicking the Add a field button.

We can convert this to lowercase using the lower function, save this, and give it a new name; however, we are going to do two steps in one, so as the output matches the preceding example, the other operation is removing the domain part of the URL; for that, I am using the "replace" command before finally naming it something and clicking save. I am going to name it Landing Path as this is a good reflection of what it is.

So now when we add it back to the report, it matches the data in Google Analytics (GA), which means it is ready to blend!

With the first data source selected (from Google Search Console), click blend data and then "add another data source"; now we get to have a table with metrics from both Google Analytics and Google Search Console – a single table with all the metrics for all the pages.

We can take this a step further and group pages, and this is where it gets really exciting on an ecommerce website – how much traffic is coming to a product page, a category page, and the blog site – suddenly we have insight into how the website is working; to do this, we can group pages using the case statement, but you need to do this to the data in advance, blends on both sides that match. See Figures 6-8 through 6-12.

Field Name

Landing Path

Field ID

calc_fn3gdkcr6b

Formula ⑦

```
1   REPLACE(LOWER( Landing Page ),"https://www.domain.com","")
```

Figure 6-8. Formula to lowercase and strip out text from the landing page

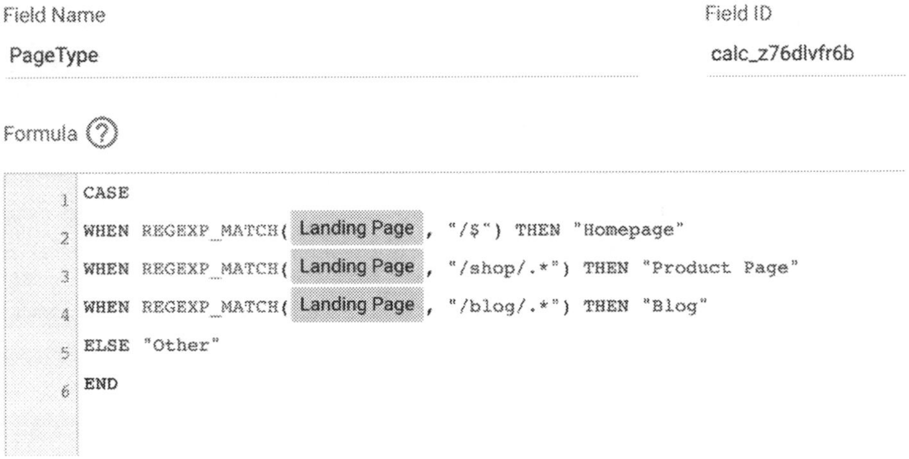

Field Name

PageType

Field ID

calc_z76dlvfr6b

Formula ⑦

```
1   CASE
2   WHEN REGEXP_MATCH( Landing Page , "/$") THEN "Homepage"
3   WHEN REGEXP_MATCH( Landing Page , "/shop/.*") THEN "Product Page"
4   WHEN REGEXP_MATCH( Landing Page , "/blog/.*") THEN "Blog"
5   ELSE "Other"
6   END
```

Figure 6-9. Using the case statement to classify pages

This is the sort of comprehensive reporting that marketing managers love, granular enough that you can get insights but top level enough that you can see trends and movements.

The case statement is an incredible way of retrospectively grouping content to analyze and see trends, as it can use a regular expression as it is a more powerful way than other simpler matches.

How to Use Calculated Metrics in Blends

I (Gerry) once worked on a project where we were unsure about the accuracy of Google after a migration. We wanted to compare the organic traffic to the clicks, and so we followed the steps of combining the clicks to the organic traffic – a graph of Clicks from Google Search Console and a graph of Google Organic traffic (created by adding a filter to a graph of sessions). See Figure 6-10.

Figure 6-10. Applying a filter to a graph

We selected both graphs and blended. The graph that came out was perfect, but we wanted to be able to calculate the difference between the two – so we add in other metrics. See Figure 6-11.

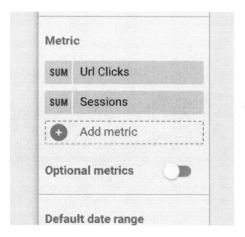

Figure 6-11. Adding additional metrics

Adding metrics gives us the ability to include a new calculated metric based on these two. See Figure 6-12.

Figure 6-12. Calculating a metric within a blend

And suddenly we get a running relationship between the clicks into the website and the data – ideally if you have filtered to just organic, it typically lies within the range of 0.5 and 1.5 and is something I often test when auditing websites. As it is a ratio, it is important to set the aggregation to "Average"; else it might start to become incorrect when displayed later.

Using the average number in the graph, I usually set it to being on the right-hand side while still displaying the other two numbers and making it bolder; this can be done by clicking on the graph and clicking into the style tab of the configuration.

The opportunities with blended data are only really limited by your imagination – it is always worth making sure that the blends are accurate and that nothing is being aggregated or you are doing anything that delivers incorrect results but discovering sources of data and the impact on others is a marketing analyst's dream.

Recap

Blending of data is initially quite daunting, but by following the steps and knowing what you want to achieve, hopefully you can see that it is a good option which isn't as challenging as it would appear on the outside. It is powerful and can deliver to you insights that wouldn't normally be available.

Remember that the key must match up and that the data must be able to match using this key and everything else will fall into place. Blending two charts is the easiest way to get started, but from there it can be increasingly more intricate.

Custom Data Visualizations

Over the course of the previous chapters, we have been showing you how to connect your business data to Data Studio and make it actionable. The challenge is that despite Data Studio being a superbly flexible tool, there are always going to be use cases that it's not able to solve for. This is due to the fact that if Data Studio was to cover everything, it would reach the point where it starts to become bloated and harder to maintain (as well as needing an army of developers to keep it going)!

The aim for most data products, as well as Data Studio, is simply to cover the majority of use cases in an easy way, and create frameworks to allow others to extend Data Studio to answer their own use cases.

Typically, these "extensions" would require someone who knows how to use JavaScript, in order to build out this functionality. When it comes to actually using these extensions in your reports, you don't need to know any programming languages, and they will function in exactly the same way as your other elements in Data Studio.

© Grant Kemp, Gerry White 2021
G. Kemp and G. White, *Google Data Studio for Beginners*,
https://doi.org/10.1007/978-1-4842-5156-0_7

What Customizations Does Data Studio Offer?

There are two types of customizations that you can leverage if you want to extend Data Studio:

- Custom community connectors
- Custom community visualizations

Custom Community Connectors

If you find that the platform that you are using doesn't have the ability to connect into Data Studio or one of the standard file formats that it uses, then you can choose to build your own community connector.

Custom community connectors are the standard way to connect Data Studio to pretty much any data source you can think of if the readily available connectors do not suffice.

Community connectors are usually built using Apps Script which is Google's standard scripting language for building add-ons for Google's suite of productivity tools.

The scope of building a connector is actually pretty straightforward if you have used Apps Script previously. Google has provided a great deal of useful resources on their developer site (https://developers.google.com/datastudio/connector) including some premade templates.

▓ Tip If you are interested in learning more about Apps Script and seeing what else you can build with it, then you can find out more at https://developers.google.com/apps-script.

Once a Data Studio connector has been created, then it can be shared with others either by publishing it in the Data Studio Connector Gallery or by sharing a link. The Connector Gallery will require someone at Google to review and pre-approve it prior to publishing.

Custom Community Visualizations

Custom community visualizations are a newer feature than custom connectors and give you access to make your reports look beautiful and bespoke. A custom visualization can be added when you are not satisfied with any preexisting ways that will adequately display the data you are trying to show.

As an example, if I worked for a theater company, and I wanted to display critics' review scores and reviews, I would want to find a way to process the data into a final star rating visualization for my report. I could think about potentially using something like a "Scorecard" or a "bar graph" control, but it won't give me the instant understanding when compared with a user seeing several stars next to a review. If you look at Figure 7-1, you can see an example of the "Customer Reviews" visualization by Baguette Engineering which displays

- Name of the item being reviewed (in this case Movie Name)

- Star rating out of five

- Date that the review was posted

- Quote or comment from the review

Figure 7-1. "Customer Reviews" visualization by Baguette Engineering

How to Add a Custom Visualization to Your Reports

In order to create the review visualization in Figure 7-1, all we need to do is

1. Create a Google Sheets file which looks like Table 7-1.

Table 7-1. Data for the "Customer Reviews" table

Review ID	Name	Review Date	Comment	Ratings 0–5
1	Movie Name	20190512	It was a triumph	5
2	Action Movie	20190515	It was bad	2
3	Sci-fi Movie	20190516	It was great	4

2. Open the Data Studio report for the dashboard that you are working on.

3. Connect your "Customer Reviews" Google Sheets data source by going into the Resource menu ➤ Manage Added Data Sources and clicking Add a Data Source in the bottom left corner (Figure 7-2).

Figure 7-2. "Manage Added Data Sources" screen

4. Select Google Sheets Connector and select your "Customer Reviews" file.

5. Click the "Add" button to start the data flowing, then click the "Add to report" button.

6. Click the "Add Custom Visualization" button in the toolbar. The button will look like a series of shapes followed by the plus sign (Figure 7-3).

Figure 7-3. Add custom reviews button

7. The Community visualization panel will open (as seen in Figure 7-4), then select "Explore More" to see the full list of current community visualizations.

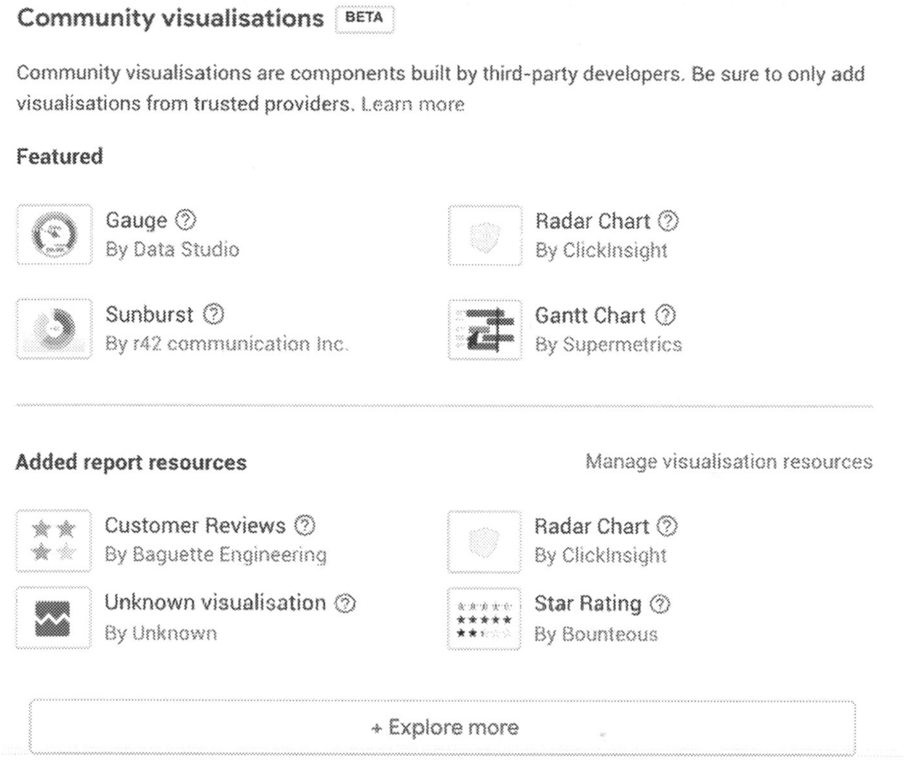

Community visualisations BETA

Community visualisations are components built by third-party developers. Be sure to only add visualisations from trusted providers. Learn more

Featured

Gauge ⑦
By Data Studio

Radar Chart ⑦
By ClickInsight

Sunburst ⑦
By r42 communication Inc.

Gantt Chart ⑦
By Supermetrics

Added report resources   Manage visualisation resources

Customer Reviews ⑦
By Baguette Engineering

Radar Chart ⑦
By ClickInsight

Unknown visualisation ⑦
By Unknown

Star Rating ⑦
By Bounteous

+ Explore more

Figure 7-4. Community Visualizations screen

8. Select "Customer Reviews" and then click and drag where you want it to appear on your report.

9. You should be seeing an error screen which says "Community Visualizations Disabled" (as shown in Figure 7-5). Don't panic, as this is expected.

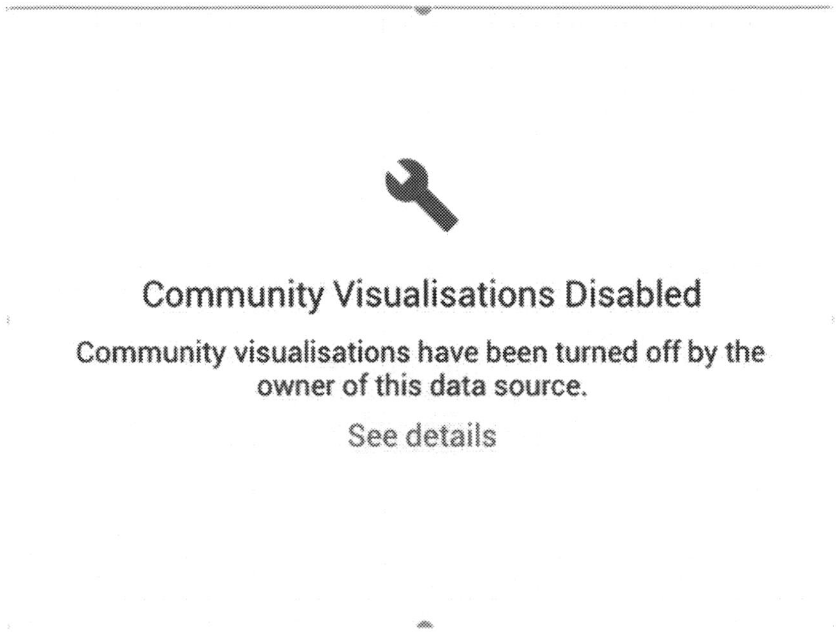

Community Visualisations Disabled

Community visualisations have been turned off by the owner of this data source.

See details

Figure 7-5. "Community Visualizations Disabled" error message is shown when you haven't allowed the use of Community Visualizations in your data source

Privacy and the "Community Visualizations Disabled" Error

The "Community Visualizations Disabled" error is a very common error that comes up regularly with users of Data Studio. The issue occurs because, by default, we didn't authorize the use of community visualizations as part of our data source setup process.

Despite the frightening term, the occurrence of this error is a reflection of a protective default setting in Data Studio. Since community visualizations may contain code from third parties, you need to consider potential risks before you enable them.

The main consideration is around assessing the level of risk to your data. You are relying on the developer of that community visualization to be trustworthy with your data and that their visualization will not generate any errors with the presentation of the data.

Google has therefore added this extra safeguard to your data sources that requires you to "opt in" to allow community visualizations. Whenever I opt in, I usually do

- A little more checking on the results of the visualization just in case there is an error.

- A little more consideration about how risky the data is. Since third-party code is involved, you should also consider the worst-case scenario that the code could stop working or could be broken by a new release.

Whenever you create a new data visualization source, you should think about whether you would want to have Community Visualization data accessing this data. For example, the movie reviews data is a low-risk data source and would be fine to pass through a Community Visualization.

After our risk assessment, we can proceed to remove the error and activate the data source for use with community visualizations, following the next few steps:

1. Start by selecting Resource ➤ Manage Added Data Sources button.

2. Click the Edit button next to the Data Source you want to modify. In the top row you will see there is a "Community Visualization access" tab.

3. Once you select the "Community Visualization access" tab, you just need to switch it to ON and click Save (as shown in Figure 7-6), then click the Finished button.

Figure 7-6. Enabling Community Visualizations access

4. You will now likely see that your custom visualization has appeared. This is not quite a time to celebrate yet as there is a bit more work to be done. If you click the custom visualization, you will notice that the dimensions and metrics probably aren't mapped to the correct fields.

5. In order for the community visualization to work, you need to make sure that the correct fields are dragged into the data fields with the same name (as shown in Figure 7-7).

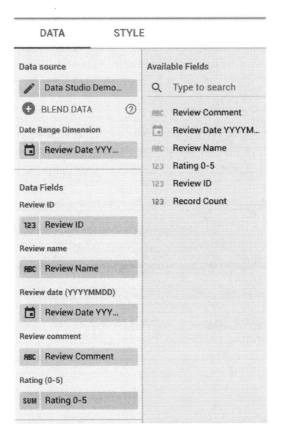

Figure 7-7. *In order for the community visualization to work, you need to make sure that the correct fields are dragged into the data fields with the same name*

6. Make sure you have a look at the "Style" tab in the Data Studio right-hand pane, and you will also see you can select which row to use as the basis for your review visualization.

That's it – you have a working community visualization working in your report. Now that you have seen how easy it is to add, let's explore some of the other options we have for custom data visualization.

Examples of Community Visualizations

Now that we have got our hands on community visualizations, let's look at a few more.

When exploring which community visualizations I can use, I generally perform the same thought process:

1. What does my data look like?

2. What impact do I want from the data report?

3. How do I make my report look visually appealing?

4. Can I find a community visualization that will do some of the hard work for me?

Step 4 always has me diving into the "3 dots" menu on each community visualization, specifically the "Learn More" option (shown in Figure 7-8).

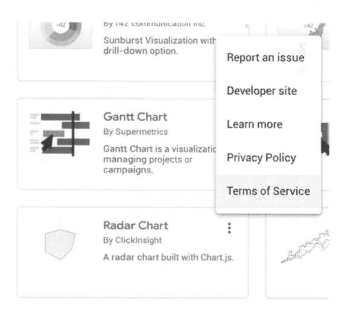

Figure 7-8. The "3 dots menu" for custom visualizations

Each of the community visualizations comes with a "demo report" which allows people to see what the visualization should look like and also check for any errors. Let's take a deeper look at another community visualization chart called the "radar chart" or what most people call it in analytics... the spider chart.

Yikes, Spiders!

The radar chart or "spider chart" is a very useful data visual when it comes to general business consulting and strategy. There is an elegant simplicity in the way that it can answer the question: "Where should we focus?"

The spider chart is made up of several equiangular lines which represent different metrics. In the spider chart in Figure 7-9, we have plotted the scores of three pupils on the chart.

In the spider diagram, we can read it as follows. Person 1 is really doing well at subjects like English, Maths, and French but is doing the poorest in the class for physical activities such as dancing and sport. The teacher could thus help them by encouraging more physical sports with the student or trying to find something that they would enjoy and get exercise.

Of all the teachers, it looks like the Maths teacher is consistently getting the better grades for all the students so should be commended on their work.

Simple and powerful.

▓ **Tip** Filters work the same way with community visualizations. Thus, if you want to break out a set of students into their own charts, then you would just need to duplicate the chart and add a filter.

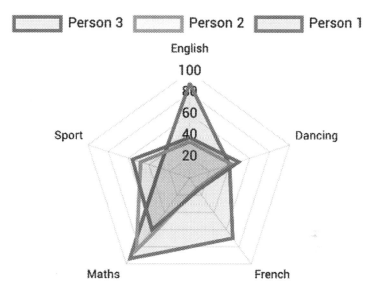

Figure 7-9. The "spider chart"

The spider chart community visualization was created using a free and open charting library called "Chart.js" which is really vibrant and well made. Additionally, it has the value of being open for anyone to use or share.

Recap

In this chapter, we have deep dived into how to make your Data Studio reports look more customized and how you integrate other data sources. We finished off by reviewing a few of the popular Data Studio Community Visualizations.

Why Stop There?

Data Studio Is Limited Only by Your Imagination

Roads? Where we're going, we don't need roads.

—Dr. Emmett Brown in *Back to the Future* (1985)

I (Gerry) am almost guilty of overusing Data Studio. When I want to visualize or analyze something, it is surprising how often I turn to Google Data Studio while auditing a client's analytics. I will create the longest possible page and use table after table to demonstrate issues with a little text around it. Using custom formatting I can highlight key issues, and from there we can monitor them to see if they have been fixed up. I will create visualizations of every campaign and will use it to present tables of data that I want clients to see, interact with, but not to edit.

Data is often interesting, often beautiful, and sometimes hypnotic. Critically it is all about interpretation. If you haven't had the opportunity to read *Freakonomics* by Stephen J. Dubner and Steven Levitt (2005), it explains why correlation isn't the same as causation, and it is definitely a recommended read on how to interpret data.

© Grant Kemp, Gerry White 2021
G. Kemp and G. White, *Google Data Studio for Beginners*,
https://doi.org/10.1007/978-1-4842-5156-0_8

Think about the incredible story of how mathematician Abraham Wald factored survivorship bias into his calculations when considering how to minimize bomber losses to enemy fire. Rather than putting more armor where there were more bullet holes, he said they should put more armor where there were fewer bullet holes. The reasoning was that the planes that were hit in these areas probably didn't make it back and therefore couldn't be counted. This is an oversimplification of the story and the mathematics for the sake of this book, but you get the idea. The point is that without the full picture, data is often meaningless. Correlation doesn't always equal causation. Sometimes it is simply random and sometimes there are other reasons.

It is amazing how many digital marketers are "channel centric." They focus only on how their own channel is performing, and when looking at drops or rises in performance, they don't look at cannibalization from other channels. Google Data Studio (GDS) gives us the ability to share subsets of data from across the business, which allows us to have a broader view. We can effectively start to see the missing pieces, and we can then really know where to armor the plane or improve the business.

When used effectively, Google Data Studio helps give us that access to the more complete picture. It is a tool that gives us far more data sources in a single place, data that normally is challenging to combine, and it gives us an accurate picture if configured correctly.

By now, we hope that you have seen how GDS allows you to retrospectively clean up data, it allows you to segment and share it, it gives more people a more complete picture, and it makes an awesome presentation layer.

In this chapter, we'll go over some interesting ways you can use Google Data Studio.

Blog Performance

The agency that I (Gerry) worked had incentivized the employees to write more blog posts by gamifying it: whoever's blog post has driven the most views by the end of the year wins. It didn't take me long to create a visualization for this incentive using Data Studio.

Considering the needs and habits of my target audience (my teammates), I chose to embed the report by clicking the Share drop-down button and selecting Embed report (Figure 8-1). I then made the dimensions of the report mobile phone size, enabled embedding, and shared the "embed" URL (Figure 8-2). This means the team can check it on their phone and see it without being on their work computer.

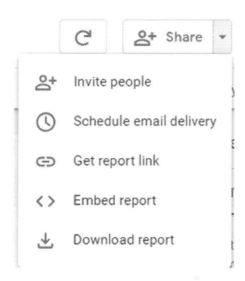

Figure 8-1. Embed report option

Embed Report

☑ Enable embedding

○ Embed Code ◉ Embed URL

Paste the following into your site:

```
<iframe width="320" height="732" src="https://datastudio.google.com/embed/reporting/83557c98-
7af5-462e-ba1b-78322a3ffac2/page/3KGTB" frameborder="0" style="border:0" allowfullscreen>
</iframe>
```

Width (px)   Height (px)
320   732

DONE   COPY TO CLIPBOARD

Figure 8-2. Embed options for a report with recommended sizes to see on a mobile

In one of the charts on my report, shown in Figure 8-3, we can clearly see that the winner at the moment is the one with the steadiest growth written on Python scripts. Sharing data on content with writers pushes them to create more and better quality. A content dashboard can show the keywords that are ranking for a selected post or the engagement with that page. I created additional tabs to track additional views of the data, partly to make sure that all the engagement metrics were set up right and we also track the author as a custom dimension into our analytics.

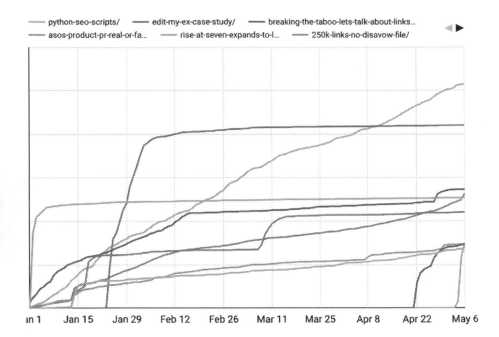

Figure 8-3. Cumulative page views for blog posts on an agency website

Creation of this widget in Data Studio has led to more colleagues producing reports and caring about the quality of the delivery. Remember, if you try this out, you can style the report however you like and embed it into web pages that change in time.

Campaign Performance Mini Reports

For every agency campaign that we do, we quickly create a campaign performance report, typically based on Google Analytics data but often shaped to pull out interactions such as video shares (including YouTube data). When the Public Relations team records coverage, it goes into a BigQuery table (it even stores screenshots) that can then be pulled through, and suddenly it becomes a report that can be easily shared with a client as well as providing internal progress reporting.

Combining, blending data from multiple sources within reports that are easily and quickly shared with clients was something we could only dream about a few years ago.

Front End for Back Office Data

Another thing about Data Studio is that it can be an awesome front end for back office data if we can pull it into Google Sheets or a database. For example, I once had to create a visualization of how many vacation days our team had already booked, had planned for the future, and had already taken. The data was already in a spreadsheet format, so it was an easy thing to pull into Data Studio. One thing I had to do was create a new sheet to clean the data. I ensured columns were correctly lined up so we could quickly see by team and then by person how much was left, taken, booked. Once I fed it into Data Studio, we magically had a report that would automatically update.

Data Studio is a versatile data visualization tool. It is not just for marketing data. I often find that friends who work in other roles or industries are pleasantly surprised when you can take their data and visualize it in Data Studio.

Embedded Forms

You could also embed a form on your Data Studio Report. For example, I wanted to embed a form on a report that asks people to recommend who I should follow on Twitter (Figure 8-4). This is a quick report that you could make in a few minutes. I created a Google Form that asks for a Twitter handle, the person's name, and why I should them. I took the URL for the Google Form, went into my Google Studio report, clicked the Embed URL button on the toolbar, placed the form on the page, and put the URL in the "External Content URL" field (Figure 8-5).

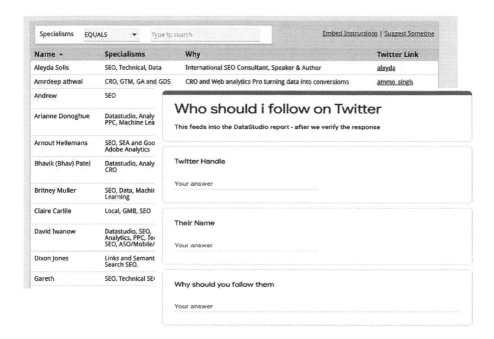

Figure 8-4. Embedding a form into a Google Data Studio report allowing you to add in more data

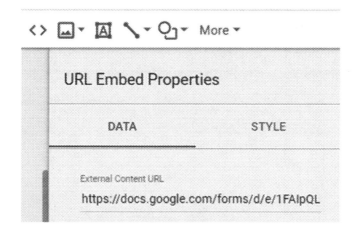

Figure 8-5. URL for an embedded form

This took minutes to build, but if I had developed it as without Data Studio, it would probably have taken a day or two to create. It is sharable and other people can embed this into their own website.

Although in this example I have shown how a form can be used to display data on a report, you can also use a form to get feedback on a report. As we mentioned earlier in this book, it's a good idea to put a footer bar with your contact details in case someone has a question or correction. You can do this by embedding a form on the last page of the report for feedback or for change requests for the report. It is a smart way of letting users submit changes and tracking them, which is a quick win.

Tips and Tricks

Data Studio has so many hacks and hidden methods that allow you to speed up things, and every few weeks an update to Data Studio brings even more. Here are a few of the best hacks I've found out there:

- **Don't reinvent the wheel** – You can get so much inspiration from templates and examples out there. You don't need to start from scratch every time if you have a good example on hand. Templates are available at datastudio.google.com/gallery. There are some amazing ones that can inspire or become your starter.

- **See if your reports are being used** – You can add an analytics code snippet to your report templates to check usage and if people are engaging with them. It's amazing how many reports are created and not used. It's also interesting to see which ones are or some are used daily. An analytics code snippet can show you which ones need cleaning up, consolidating, or removing.

- **Chart styles can be copied** – You can copy and paste (Paste special ➤ Paste style only) the styling of one chart to another so the style is consistent throughout your reports, including rounded edges, line thickness, and colors. When I found out about this, I couldn't believe how much time this would have saved me in the past!

- **Apply interactions filter to nearly everything** – Almost everything can filter everything else, which gives it some serious power to allow you to drill into what you want. If you want this to apply to just a subset, you can group them, including date filters.

- **Advanced dates** – Some clients may need to see year-on-year performance, but at a day-by-day level as opposed to date-by-date. This is generally the case when a business follows a very trend-driven calendar. As this isn't a built-in option, we have to get a little bit more technical to achieve this. Under your comparison date range selector, select "Advanced" from the list options. Once this has been done, you need to choose a fixed start date. To do this, go back to the previous calendar year and select the day that corresponds with the first date in the current year. After you've done this, it's time to set the end date. To do this, simply select "Today," "Minus" and then enter 365 as the number of days.

- **Retrospectively clean parameters from landing pages** – Though we try to have the cleanest data going in, if you are using Google Analytics, filters allow you to clean up some of the strange issues, but you can't do this retrospectively like you can in Data Studio. This is where custom fields can be used. Creating a new field in your data like this

  REGEXP_REPLACE(Landing Page, '\\?.+', '')

  strips out any of the odd parameters that social networks such as Facebook and Twitter can append. We found that by applying that to one report, some socially shared posts doubled in visits. You can also strip off and consolidate trailing slashes REGEXP_REPLACE(Page,"./","").

- **Clearly identify report metrics** – Have guides that don't just say what the metric is but what it means. You might need to add a glossary at the end of the report explaining some business metrics or simple analytics definitions such as what is a session or what the bounce rate means. On a side note, always include the y-axis label on charts, as it is frustrating trying to decipher metrics in Google Data Studio without it.

- **Use a Swatch to create a color style sheet** – If you want to style your charts and your assets quickly, you can't use CSS but you can import a graphic block that you can create in PowerPoint or some other quick graphic editor. Take a few colors from a client's website or logo to ensure that we are using their own branding throughout their reports.

Many thanks to Al Wightman, Omi Sido, Jill Quick, and Nick Wilsdon for their contributions.

There are plenty of other hacks out there, and more that haven't been discovered yet.

Web Vitals

Google, itself, uses Data Studio as a front end for some data sources now. An example of this is the Web Vitals project (https://web.dev/vitals/), which is a set of data points that show Google's view of your site. You can find out from real users experiences stored in a Google BigQuery table that they've made public and they have made it accessible through a Data Studio report which is easily configured to be focused on your site (or a site belonging to a competitor). See Figure 8-6.

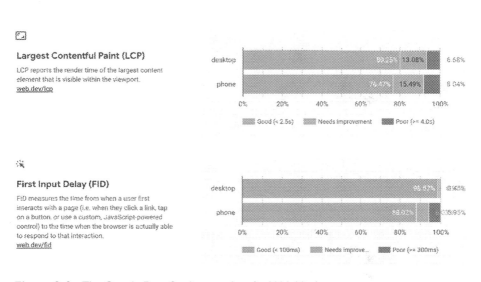

Figure 8-6. The Google Data Studio template for Web Vitals

As tools, services, and public data sets increasingly provide access to their data in Data Studio, it is clear it isn't going away any time soon.

The combination of data sources and visualizations means tracking is possible more than ever with simple visualization.

Hack in the Data Through Google Sheets

There is an increasing number of connectors out there now. Despite that, within the marketing world, there are always data sources that aren't easily available. For this reason, you can connect to these data sources, such as IFTTT (If This Then That), by connecting to Google Sheets (Figure 8-7).

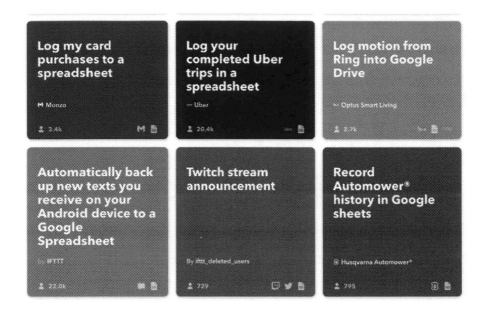

Figure 8-7. *IFTTT samples that can be used to pull data into Sheets*

You can track tweets about conferences (if they have a hashtag or common name) straight into a Google Sheet, and from there we can see the types, frequency, and who is tweeting the most in an almost live Data Studio report.

We can also track how often people complete tasks in Data Studio by hooking our project management tool to Google Sheets through Asana – a total view of how a business operates results in informed leadership delivering better results.

A popular task within any company is to look at the team allocation, how much time is dedicated to a client vs. how much they are paying, who is overworked, and who is underutilized. While it is possible to do much of this in a spreadsheet, by using the APIs within Data Studio gives you a live dashboard that gives you the ability to look quickly and efficiently at status. If you can look at this dashboard live as we go along, it give you the ability to effectively slot in new projects and client requests or compensating for an underperforming team member.

Recap

Ultimately, Google Data Studio makes data that is often locked in silos more accessible. If it was once hidden in a spreadsheet or unavailable in a datastore, it is now accessible, throughout businesses and even sometimes publicly available online.

And of course, if you have come to the end of this book and you've built something inspirational or entertaining, feel free to share it with us on Twitter!

Index

© Grant Kemp, Gerry White 2021
G. Kemp and G. White, *Google Data Studio for Beginners*,
https://doi.org/10.1007/978-1-4842-5156-0

Printed in the United States
By Bookmasters